PHILOSOPHY OF NURSING

Other interview books from Automatic Press ♦ V/I P

Formal Philosophy

edited by Vincent F. Hendricks & John Symons November 2005

Masses of Formal Philosophy

edited by Vincent F. Hendricks & John Symons October 2006

Intellectual History: 5 Questions

edited by Morten Haugaard Jeppesen, Frederik Stjernfelt & Mikkel Thorup October 2006

edited by Morten Ebbe Juul Nielsen December 2006

Philosophy of Technology: 5 Questions

edited by Jan-Kyrre Berg Olsen & Evan Selinger February 2007

Game Theory: 5 Questions

edited by Vincent F. Hendricks & Pelle Guldborg Hansen April 2007

Philosophy of Mathematics: 5 Questions

edited by Vincent F. Hendricks & Hannes Leitgeb January 2008

Philosophy of Computing and Information: 5 Questions

edited by Luciano Floridi Sepetmber 2008

Epistemology: 5 Questions

edited by Vincent F. Hendricks & Duncan Pritchard September 2008

Philosophy of Medicine: 5 Questions

edited by J. K. B. O. Friis, P. Rossel & M. S. Norup September 2011

Narrative Theories and Poetics: 5 Questions

edited by Peer F. Bundbaard, Henrik Skov Nielsen & Frederik Stjernfelt 2012

Philosophical Practice: 5 Questions

edited by Jeanette Bresson Ladegaard Knox &

Jan Kyrre Berg Olsen Friis 2013

See all published and forthcoming books in the 5 Questions series at www. vince-inc. com

PHILOSOPHY OF NURSING
5 QUESTIONS

EDITED BY

ANETTE FORSS
CHRISTINE CECI
JOHN S DRUMMOND

Automatic Press ♦ VIP

Automatic Press ♦ VIP

Information on this title: www. vince-inc. com

First published 2013

Printed in the United States of America
and the United Kingdom

ISBN-10 / 87-92130-49-6
ISBN-13 / 978-87-92130-49-5

Cover design by Vincent F. Hendricks

Contents

Preface

We want to begin with Umberto Eco's remark that "books always speak of other books"[1] (Eco 2004). There is perhaps a sense in which the chapters in this book will talk to one another before talking to other books. Their murmurings may light a spark that thinks the future. It would be a brazen lie to claim that philosophy of nursing is not a recently emerged discipline. It has emerged from the mists of history like a refugee. Yet it is here. Indeed we would like to think that it has been here for quite a while and that those in this collection of the 5 Questions just happen to be the ones who are currently speaking.

We do not speak with one voice. As with scholars in other disciplines who bring sustained philosophical questioning and analysis to their endeavours, nurse philosophers or philosophical thinkers in nursing, demonstrate the breadth of possibilities available to think through present issues in care, health, disease, death and vulnerability, along with the socio-political arrangements that are instituted to deal with these. Philosophy of nursing is first of all, and most effectively, provoked by the problems with which we are confronted – these problems are multiple and serious and probably not fully resolvable – but still we must try.

These problems need to be articulated, to be made sense of in ways that acknowledge the multidimensional nature of nursing as a professional and human practice in society. It is clear therefore that actual practice, the ideas that drive it and the policies and politics that seek to govern it are inextricably related. In addition, and of equal importance, there are pivotal issues around nursing's relation to science, and of course research, knowledge and education, and how these are organized and thought through from one generation to the next.

So neither do we speak of only one thing. Nursing is not a singularity that can be captured in the concept of 'nursing care' even if that is the bottom line. What you will find in these chapters is diversity, both of focus and argument across a range of issues and practices. You will find conflict, concord, a wonderful confusion, closely argued analysis, and, from one contributor, an invitation to a reckless philosophy. We make no apologies for that. But one of the things that draws this collection together is a concern with nursing as a vital human practice. Thus these interviews are threaded through with ideas of return, recuperation, and how to use the history of thinking in our discipline. The need to as-

[1] Eco, U. (2004) Name of the Rose (PostScript). London, Vintage Classics, Random House

semble, as one contributor suggests, "an *effective* philosophical bridge over which the history of ideas in nursing can be conveyed and, most importantly, engaged with critically."

In the many voices gathered here, perhaps we can also see more clearly and put to rest what may have been a question for some: doing nursing, being a nurse, contains as much a philosophical dimension as it does a practical one. Indeed, those speaking through this book make a compelling case that nurses, patients, families, communities and societies are not well served by rigid distinctions between practical and philosophical matters. In fact, there is an undeniable violence in insisting on such distinctions.

To expand upon this, philosophy of nursing does not have, or necessarily seek to establish a 'canon' as such. This is a different way of saying that what we do seek to establish is that a distinction be drawn between philosophy of nursing as a practice and the various and often 'cultish' ideologies that at times tend to cluster around the concept of nursing in ways that are largely epiphenomenal but, in extreme cases, also potentially damaging with respect to nursing's relation to science and research. This is not to say that, ethically, the relation between the various ideologies of nursing and philosophy of nursing is always one of mutual exclusion. But it is to say that the relation is not ever one of identity. This is a distinction of which all the contributors to this volume are aware. This leads us to say that the field of potential candidates for inclusion in this collection was extensive – both a boon and a curse for ourselves as editors. The list of contributors is not then exhaustive or exclusive but it is interesting. Indeed contributors to this book reflect a variety of approaches, interests and commitments, a current representation of nursing's well known 'eclecticism', a tradition that can be fruitful but that can (to revisit our opening remarks above) also render it all the more important that the chapters in this book talk to one another, and also talk to other books. We also suggest that an unanticipated outcome of bringing these voices together in this way is to develop a better appreciation of what moves people – we want a diverse field of scholarship but not a polarized or rigid one – and we can also begin to grasp, in the midst of this diversity, a blended coherence of the field. We are thus most grateful to all of our contributors for their thoughtfulness, for their insights and their analyses, and most of all for their sustained concern for nursing in all its aspects, for nurses themselves, and for those whose need is for nursing care.

Stockholm (Paris), Edmonton & Dundee, June 2013

Anette Forss,
Christine Ceci &
John S Drummond

Acknowledgements

"Cover photo, statue of Hygeia, Greek Goddess of Health, designed by the Manchester Sculptor, John Cassidy in 1897. The statue stands in the public gardens of Duthie Park, Aberdeen. (See: http://www. johncassidy. org. uk/cassidy2a. html) We wish to acknowledge the co-operation and support of Charlie Hulme for sourcing the image of Hygeia. Thanks also go to Steven McRae of Aberdeen, Scotland, for the original photograph, reproduced here with permission."

We are particularly grateful to the contributors for devoting time to writing such erudite, enlightening, and often thought-provoking interviews, and grateful to the philosophical community in general for showing interest in this project. In addition, we would like to thank editor-in-chief Vincent F. Hendricks and associate editor Henrik Boensvang of Automatic Press ♦ VIP.

Stockholm (Paris), Edmonton & Dundee, June 2013

Anette Forss,
Christine Ceci &
John S Drummond

1

David Allen

Professor, Psychosocial & Community Health
University of Washington, Seattle, WA, USA

---

1. How were you initially drawn to philosophical issues regarding nursing?

I came into nursing with two perspectives that led me in a philosophical direction. I had three degrees in the humanities (English—with a criticism emphasis; Theatre—again with a focus on critical and performance theories-- and Philosophy/aesthetics). These fields take language as central but particularly in graduate school, I worked in American and British analytic philosophy. My dissertation tried to interrogate French phenomenology (in aesthetics) through analytic philosophy (such temerity). But I am also an un-apologetic 'child of the 60s' and was brought up in second wave feminism, civil rights, union and anti-war struggles. These movements highlighted the political effects of language even as they seriously underestimated them.

Entering nursing in my early 30s immediately confronted me with two profound puzzles. First—and I mean this literally—I could not understand nursing theory. The sentences simply made no sense and the style of argument was completely foreign. The second puzzle was organized nursing and nursing education's largely negative orientation to both feminism and labor rights. Just as I could not understand nursing theory, I was befuddled by how an institution run by and for women, a field in which women were deeply underpaid, could fail to align itself with feminist politics. There seemed to be a strange nominalism: that if you called something a profession, you somehow became one (leaving aside the assumption that that's a good goal and, of course, ignoring the fact that medicine achieved 'professional' status through white, masculine privilege and political struggle, not through science and service).

Consequently, much of my career has been focused on language in nursing and on the relationship of language use to political and ethical interests. Historically, I was pretty fortunate in that voices like Peggy Chinn, Barbara Bowers and Jan Thompson had started to emerge plus I was able to get an appointment in Women's Studies when I joined the nursing faculty at the University of Wisconsin-Madison. Although I occasionally thought of myself (egotistically) as a lone voice in the wilderness, having even a small set of colleagues to be in conversation with is invaluable. For scholars like Evelyn Barbee who were trying to make an argument about racism in nursing, it must have been much more difficult. I was pretty blind to my whiteness at the time, but my own gender was pretty salient. The dean who initially hired me cited it as a major asset while an associate dean who voted against my tenure (in open meetings) declared it undesirable.

2. What, in your view, are the most interesting, important, or pressing problems in contemporary philosophy of nursing?

I think the greatest threat to nursing on many levels, not just theoretical, is its intellectual isolation and self-referential tendencies. To offer just one example, it is common to read nursing literature—including philosophical nursing—and find almost exclusively nursing citations in the references. When I entered nursing (in the late 1970s), the philosophy of science citations were already twenty years out of date. Scholars were (and still are) using 'positivism,' 'empiricism' and 'quantitative' as if they are synonyms. That's not just embarrassing; it's profoundly misleading and historically incorrect.

Apart from intellectual isolation, there is a deeper isolation that comes from being uncritically embedded in western medicine. When I arrived at the University of Washington in 1988, the percentage of international students in our doctoral program was increasing. My mentor, Professor Oliver Osborne, stood up in a meeting and said if we weren't prepared to include content on the role that western medicine played in the colonization of their countries, we shouldn't admit the students. When he sat down, there was a pause and then the meeting continued as if he'd spoken in ancient Hebrew. In contemporary parlance, the fact that explication and critiques of neoliberalism are rarely addressed in our literature or our courses leaves students disempowered in terms of how they are positioned as "individuals" in global capitalism, much less in how they might relate to fields like "global health."

I need to recognize here that this is also an effect of disciplinary structure: When I began teaching in nursing, my colleagues had their degrees in sociology, psychology, pathophysiology, etc. But when one is producing nursing PhDs one is pressured to hire nursing PhDs. The interdisciplinary grounding of our field is increasingly homogenous. Look at schools of education in the US for how that plays out historically.

A third threat to our field is the absence, indeed the lively avoidance, of critique. In most other fields in which I work, a portion of every paper and indeed sometimes entire books and journal issues are devoted to criticizing other's work.

Relatedly, a pressing need is to skill up our students (and ourselves) in argumentation—verbal as well as written. For several years I taught the undergraduate nursing research course and made the arguments that (a) research is not (just) the reporting of findings but more seriously a form of argument with assumed norms, audiences, etc. and (b) most novice (and beyond) nurses would be better served by learning how to hold their own in verbal debates about practice than in learning research methodology they will rarely use. These aren't separable, of course. I'm creating a rhetorical binary because many of our oral arguments use research findings as evidence.

3. What, if any, practical and/or socio-political obligations follow from studying nursing from a philosophical perspective?

For me, the primary obligations are ethical and educational. While there are important exceptions, human beings become persons—people, subjects, citizens—largely through the acquisition and use of language. From Marx to Freire to Gramsci to Chomsky, philosophers have argued that the best way to create a docile citizenry is to shape and control their acquisition of language. We make the world through the word. So for me, the first ethical obligation is to try and explicate what sort of a world we're building within our language. Who gains, who suffers, who heals, who is lost.

Consequently, for me the heart of education is creating alternative vocabularies, helping students and ourselves envision and even create better worlds through better words (representations, narratives, etc.). I profoundly disagree with the neoliberal practice of universalizing the "market" as a model for social reality. Thatcher was wrong when she said 'There Is No Alternative." There not only is one, there are many.

So, for example, I argue that the goal of our health care system is neither health, nor care, but profit. Providing care is simply a means to that end. Otherwise, having known for two generations that it is both much cheaper and vastly more humane to provide prenatal care than build NICUs, that's what we'd do. But the former doesn't generate much profit and the latter does (including good jobs for nurses).

The ethics of philosophical analysis, then, are inseparable from education. When we help our students discover and practice alternative narratives (e. g. non Cartesian understandings of the mind; non-market understandings of 'choice') we make it possible for them to make choices about how they position themselves in the world, about what sort of world they'd like to be part of.

4. In what ways does your work seek to contribute to philosophy of nursing?

Honestly, I don't think of it that way. Maybe because there was no such identifiable genre of nursing literature when I entered the field (except nursing theory, which for the most part I found extremely un-philosophical). More basically, I don't orient myself to disciplines or institutions. Since I earn my salary in them, I try to be useful, productive, of course. I try (too often unsuccessfully) to keep my focus on values and people rather than professions or disciplines or institutions. Again, this is over-reaching because professions, disciplines and institutions are the effects of people's performances. In the final analysis, though, I dream of contributing to the capacity of people to imagine and help build preferable worlds. I only wish I were better at it.

5. Where do you see the field of philosophy of nursing to be headed, including the prospects for progress regarding the issues you take to be most important?

I'm not optimistic but neither am I cynical. The hegemony of neoliberalism and its agenda for corporatizing the university is very challenging. The evisceration of public health is almost complete. It has been very successful in appearing 'natural' and inevitable, rather than the consequence of elite politics. But as I write this, students in Chile are in the streets, revolting against the state's lack of support and ideological interventions into education.

Selected Works

Allen, D.

1986."Professionalism, Gender Segregation of Labor and the Control of Nursing." *Women and Politics* 6(3): 1-24. An earlier version was published in S. Geiger (Ed.) (1984), *Sex/gender Division of Labor: Feminist Perspectives*. Minneapolis: University of Minnesota, Center for Advanced Feminist Studies.

1986."Nursing and Oppression: A Feminist Analysis of Representations of "the family" in Nursing Textbooks. *Feminist Teacher* 2: 15-20.

1987."Critical Social Theory as a Model for Analyzing Ethical Issues in Family and Community Health." *Family and Community Health* 10(1): 63-72.

1987."The Social Policy Statement: A Reappraisal." *Advances in Nursing Science* 10(1): 39-48.

1995."Hermeneutics: Philosophical Traditions and Nursing Practice Research." *Nursing Science Quarterly* 8(4): 174-182.

1996."Knowledge, Politics, Culture, and Gender: A Discourse Perspective." *Canadian Journal of Nursing Research*, 28(1): 95-102.

Allen, D. G. and P. Hardin. 2001. "Discourse Analysis and the Epidemiology of Nursing." *Nursing Philosophy* 2: 162-76.

Allen, D. G. 2002. "September 11th and the Eminent Practicality of Post-structuralism." *Nursing Inquiry* 9: 1-2.

Allen, D. G. and K. Cloyes. 2005. The Language of Experience in Nursing Research. *Nursing Inquiry* 12: 98-105

DiAngelo, R. and D. G. Allen. 2006. """My Feelings Are Not About You": Personal Experience as a Move of Whiteness." *InterActions: UCLA Journal of Education and Information Studies*. Vol 2, Issue 2, Article 2. http://repositories. cdlib. org/gseis/interactions/vol2/iss2/art2.

Allen, D. G. 2006. Whiteness and Difference in Nursing. *Nursing Philosophy* 7: 65-78.

Bonomi, A., Allen, D. G. and V. Holt. 2006. "Conversational Silence, Coercion, Equality: The Role of Language in Who Gets Identified as Abused." *Social Science and Medicine* 62: 2258-2266.

2

Peter Allmark

Senior Lecturer

Sheffield Hallam University, UK

---

1. How were you initially drawn to philosophical issues regarding nursing?

As a child I was a day-dreamer interested particularly in astronomy. This evolved into an interest in philosophy, partly inspired by the image of the philosopher, Mathieu, portrayed in a television adaptation of Sartre's *Roads to Freedom* trilogy; it looked cool. On television also was a series of interviews with philosophers conducted by Bryan Magee, *Men of Ideas* (Magee 1978). I still have the hardback book of the series, bought for me by sister in the hope that it would satisfy my interest before I ended up choosing philosophy for a degree course. No such luck: although I began a BA in Economics I soon transferred to philosophy. I was not particularly industrious but I was taken with Popper following a term of one-to-one seminars with one of his acolytes, Jerzy Giedymin. I also spent (too) much time reading Marx and demonstrating.

My left-wing views were in part behind my drift away from philosophy on graduating. Without overdoing my sense of purpose, which was weak, I thought it better to enter work than carry on in academia. From 1983-1989 I read little philosophy. I returned to it when I wrote an essay on ethics for a course on coronary-care nursing. This was then accepted by the *Journal of Advanced Nursing* and I was inspired to take it further, signing up for the MA in Health Care Ethics at Leeds, led by the brilliant marital partnership of Jennifer Jackson and Christopher Coope.

2. What, in your view, are the most interesting, important, or pressing problems in contemporary philosophy of nursing?

Nursing has two branches, research and practice, with the latter subdividing into education, management and clinical practice. Philosophy's place is either as an element of the theory behind research or practice. When do we need it? I am fond of Neurath's analogy.

Neurath has likened science to a boat which, if we are to rebuild it, we must rebuild plank by plank while staying afloat in it. The philosopher and the scientist are in the same boat. Our boat stays afloat because at each alteration we keep the bulk of it intact as a going concern. (Quine 1960, 3-4)

Nurses are also in this boat. Some of its planks are philosophical ideas. For example, when nurses give a drug they do so in the belief that it will have effects; but the philosophical notions of cause and effect inherent here rarely need scrutiny. But some of the planks are wonky and reliance on them can lead to our boat going astray. Take the concept of nature: it might be variously used, as in i) natural childbirth, ii) letting nature take its course and iii) forbidding postmenopausal women unnaturally to conceive using IVF. There is no clear link between each use; further, each is problematic such that reasoning based upon it could be challenged. For example, a nurse who allows a patient to die using the second sense of nature might be asked why she nurses at all: why not always let nature take its course instead? The wonky planks causing difficulties constitute the pressing problems of philosophy of nursing.

In truth, many, perhaps most, are pressing problems of philosophy of health care practice, problems for other health practitioners as well as nurses. Those that are linked particularly with nursing seem often to be the product of an undue desire to define nursing as different from and independent of other practitioners. Two examples with which I've been academically involved are the notion of nurses acting as patient advocates and the notion of nursing as having an ethics of care that is importantly different from the doctors' ethics of justice (Allmark and Klarzynski 1992; Allmark 1995). Even here, nurses are not alone in the error. And I'm not sure how pressing either problem is. Nurses who believe they are acting as patients' advocates seem likely to be good nurses doing what is best for the patient; they are simply misdescribing it. The same is true *mutatis mutandis* of nurses acting in line with an ethics of care. Thus these seem to be wonky planks that don't have adverse effects on the Good Ship Nursing. What might follow is that those of us who engage with issues in nurse philosophy should ask first whether anything in nursing research or practice is at stake. I wonder whether, had I done this, I would have avoided the advocacy and care-ethics debates. So, what are the pressing problems, those that make a difference?

The problems will not necessarily be clear to us; if they were clearly problematic then they would not be our underpinning beliefs. To illustrate by historical analogy, we find it hard to understand how in the UK until fairly recently apparently reasonable people could not see the problem in slavery, or the miserable conditions of workers, or in the starvation of the Irish population. However, it is instructive to examine how these inequities were rationalised at the time for, as I shall argue in the next section, similar rationalisations persist in some present justifications for inequalities in wealth and health. As such, I aim to show that these inequalities are one of the pressing problems in philosophy of nursing and health care. Our philosophical perspective will affect our attitude to health inequalities and our action to tackle them.

3. What, if any, practical and/or socio-political obligations follow from studying nursing from a philosophical perspective?

One of Dickens' targets in *Oliver Twist* was the Poor Law Amendment Act [PLAA] of 1834. This set up a workhouse system based on two principles: first, that workhouse conditions must be worse than anything available outside the workhouse and, second, that poor relief should be available only via the workhouse. As Pope puts it, the commissioners who wrote the report leading to the PLAA,

> Sought to target those who would 'avail themselves of the mischievous ambiguity of the word *poor'* [citing Poor Law Report of 1834] namely the able-bodied unemployed, or 'undeserving poor' and punish them by a suspension of their rights. They were to be segregated both spatially, by enclosure in the workhouse and morally, through a new taxonomy that would seek to stratify society. (Pope n. d, 1)

Underlying these conclusions were various elements of political and moral philosophy, including beliefs that 1) the operation of the market would ensure fair outcomes, that is, those which reward people for their contribution to society, 2) people would always choose the most pleasurable option (e. g. leisure over work), 3) humans tended to multiply beyond the capacity of the economy to support them, 4) wages will always tend down towards subsistence level, 5) both in nature and society a system of survival of the fittest operates; associated with this is the idea that human beings are individuals in competition with one another. I shall term these the 'five Victorian beliefs' - VBs for short. The sources for some are obvious and include Adam Smith for the first, Malthus, the third, and Darwin the fifth. In Darwin's case the doctrine is dubiously attributed to him. The other two are less obvious. The fourth may be due to Ricardo or Lassalle. The second seems to have origins in Bentham's utilitarianism. Note here that ideas can be usurped by those hostile to their originators. For example, whilst the belief that people prefer the most pleasurable option is reasonably attributed to Bentham, its use to justify the removal of protection for the poor runs against his political radicalism and that of his followers, particularly, JS Mill.

Let us now turn to the persistence of the VBs, the way they are still used to explain and justify poverty. The most obvious is the second: that people choose pleasurable options; this is seen in UK newspaper stories of benefit fraud and benefit choice; it is said that living on welfare benefits instead of working is a lifestyle choice for the 'undeserving poor'; people choose to live on generous welfare benefits rather than work. The belief in the operation of the market is shown in arguments that employers cannot compete in markets whilst hampered by over-generous welfare payments and rights. Similarly, we see echoes of: Malthus in the argument that the economy cannot afford to encourage non-working parents to have further children; and Lassalle, in arguments that high wages or welfare benefits in the UK make it impossible to compete in world markets - successful countries and companies are those that pay the least to produce the most. Social Darwinism is combined with free-market philosophy to justify wealth inequality: roughly, the unfettered market produces the best outcomes for all, whatever their starting point; but the operation of the market will inevitably produce economic inequality because people differ in their talents and willingness to work.

These five VBs manifest in a particular approach to health inequality, one that seeks to reduce such inequality by changing the behaviour of those at the wrong end of it. In the first place, insofar as poverty is due to the poor themselves (as it is with the 'undeserving poor'), so are its consequences, such as ill health. Further, in line with Social Darwinism and the Bentham 'pleasure principle' the poor are, for some reason, particularly likely to make choices for pleasure now that risk their health in the future: smoking, bad diets, lack of exercise, not breast feeding, and so on. If the State has a role it is to encourage the poor to change and by so doing to improve their own health - this will leave economic inequality untouched, the five Victorian beliefs decree this must be so, but you will no longer have health inequality. Inequality in people's health based on inequality in their wealth is well documented; the poor are sicker than the wealthy. Thinking based on the five VBs leads towards targeting the behaviour of the poor rather than their poverty.

This gives us a broad outline of the type of philosophical beliefs underpinning the policies that tackle health inequality by targeting the poor; what beliefs would underpin the alternative approach, targeting poverty itself? The five VBs are contentious; one source of alternatives is the Aristotelian and Marxist influenced work of Sen and Nussbaum termed, the capability (or capabilities) approach (Nussbaum 2006; Nussbaum and Sen 1993). This had its origins in work by Sen challenging the Malthusian idea that famine is a product of shortage. Sen examined past famines and found that this was almost never the case;

in famine, people starve because they are not entitled to available food not because there is no food available. Both Sen and Nussbaum also challenge the first VB, the meritocratic view of fairness. Meritocracy is compatible with great inequality in which many people lack the capability to be and to do things that are important in a good human life. The famine example is the extreme example, there is enough food but people starve. But a society is also unfair where, for example, there are enough resources for people to spend time with their family but some poor families cannot due to the demands of work.

We have then at least two approaches to health inequality based on different philosophical ideas; the individualist approach and the Aristotelian approach. We can now return to the question that opened this section, what, if any, practical and/or socio-political obligations follow from studying nursing from a philosophical perspective?

In almost all areas of health care practice in the UK, including nursing and health visiting, health inequality intrudes; individual health correlates closely to people's wealth. This is generally agreed to be a problem that should be tackled; for example, the Marmot review on health inequality was welcomed by politicians across the political spectrum in the UK (Marmot 2010). However, there is disagreement over whether the ill-health of the poor is due to poverty or to something about the poor themselves. Those who believe the latter point turn to the five VBs as justification of financial inequality and as partial explanation of health inequality. We see the latter when commentators and politicians point to the choices poor people make, with higher levels of smoking, obesity, lack of exercise, and so on. By contrast, those who stress poverty as a cause of illness suggest, for example, that the poor lack resources required for good health. This 'poverty' case faces a problem, however. This is that poverty has long been defined in relative terms; you are poor if you have a level of wealth well below the national average. But of course, if the national average continually rises then the poor remain relatively poor but in absolute terms become better off. How can this cause health inequality? As Le Fanu puts it, "Absolute poverty in the form of an inadequate diet, overcrowding, poor hygiene and lack of protection from the elements can harm the human organism and cause disease. Relative poverty cannot." (Le Fanu 2011, 371). Thus it is argued that the poor are unhealthy because of lifestyle or cultural factors, not because they are poor. They have, for example, lack of education or a poverty of ambition. At this point, though, the blame-the-poor argument runs into problems. Why should a whole group of society, which happens to be poor, coincidentally lack education or ambition or belong to cultural milieus in which unhealthy activity is the norm? Remember, those taking this position need to show that none of this is a function of poverty itself. One way this might be argued is to say that the nature of the poor has changed with upward mobility such that the poor now are also more likely to be, for example, less intelligent or to have other damaging characteristics (Mackenbach 2012). This argument requires two stages of proof: first, it must show that the characteristic is true of the poor and, second, it must show that the characteristic leads to ill health. To my knowledge, this has not yet been done; I also think it unlikely to be true.

I turn next to those who blame poverty for health inequality. They are faced with the problem of relative poverty. There are various ways they might respond. The first is that *contra* Le Fanu's point the poor in prosperous Western countries lack the resources to maintain good health, they lack the capability for health (Lynch et al. 2000). One visit to the website of the Joseph Rowntree Foundation is enough to show this point is plausible (http://www. jrf. org. uk/). The second is that inequality itself causes health and other problems; this is the line taken by Wilkinson and Pickett whose book, *The Spirit Level*, offers datasets that show numerous ways in which unequal societies do worse then equal ones across many indicators, including health (Wilkinson and Pickett 2008). The third point is that if health is (in part) a function of wealth then the wealthy will generally be healthier than the poor. No matter how wealthy a society gets, if it is unequal in wealth it will have inequalities of health. These three points do not sit together entirely easily. My belief is that the capabilities approach points towards the first explanation as most important such that in an unequal society where the poor were above a threshold level of wealth, health inequality would largely disappear. For my purposes here, though, I shall not try to argue this case. My main conclusion is that the philosophical position taken on beliefs such as the five VBs will influence the policy and practice of those dealing with health inequality.

It is tempting to think that approaches to health inequality that focus on individual change, such as stopping smoking, and those that focus on social change, such as Marmot's recommendation of a living wage for all, are complementary. The discussion here suggests them to be rooted in philosophically incompatible beliefs. And despite the oft-repeated concern about health inequality from governing parties, almost all the policies that have emerged have been of the type that focuses on the individual rather than poverty. Nurses and health visitors end up thus, for example, trying to get people to stop smoking, reduce their weight and so on, without locating this behaviour in poverty. With Dickens we are appalled by the conditions of the workhouse, but will future generations look back on the continuing influence of the five Victorian beliefs on our attitudes to health inequality with similar disgust?

To sum up, I have suggested that the health inequality is one of the important problems in contemporary philosophy of nursing. This is because the philosophical position we take on, for example, the five VBs will affect our practice as nurses. If we accept the five VBs we will continue with practice that seeks to change individuals' behaviour without changing the context in which that behaviour emerges. If we reject the VBs and accept instead something like the capabilities approach then our focus will change to look at the context as the primary influence on, for example, smoking or obesity.

Someone might suggest that nurses and health visitors are only in a position to address individual behaviour; we cannot change the world *qua* nurses although we might seek to do so as civilians. This seems correct. As health carers we see individuals and families with problems; we cannot help an individual who wishes to lose weight or stop smoking by getting them out of poverty. However, by locating the issue in inequality rather than individuals we lose the blaming mindset; we see the individual's problems as primarily political or social in origin rather than a product of moral weakness. This change in attitude is likely to have a beneficial knock-on in care anyway. More importantly, there might be small changes we can make as professionals; for example, our care aimed at helping someone stop smoking could be combined with helping them tackle their debt or housing problem.

4. In what ways does your work seek to contribute to philosophy of nursing?

In the discussion so far I have tried to show that health care policy and practice that blames individuals for their ill-health and targets them for behaviour change has origins in philosophical beliefs that are longstanding, widely held but subject to challenge. One contribution I would like to make to the philosophy of nursing is to challenge them and present an alternative. At present (2012) in the UK this seems to be swimming against the tide, the current Government now seems wholeheartedly to have adopted a belief set in which the poor are to blame for their poverty and ill health. This, of course, makes the challenge more important and I view my contribution as, at best, part of a wider movement. Given that one philosopher states that the average philosophical paper published in a peer-reviewed journal is read by seven people we should not get self-important (Crisp 2010).

Insofar as I believe that philosophical positions lead to political obligations it follows that philosophy has, or should have, effects on our beliefs and actions. This is in contrast to Wittgenstein's belief that philosophy leaves the world as it is but is in line with Marx's epigram about the point of philosophy being to change the world. However, I don't think I am Marxist these days. When Marx was voted the favourite philosopher by listeners to the BBC Radio 4 programme, *In Our Time*, my thought was that he wasn't even the best philosopher called Karl.

Karl Popper is an obvious rival. Popper's work began from his disagreement with the logical positivists. This set him at odds philosophically with Wittgenstein who was, at one time, logical positivist in all but name. Related to this was a disagreement over ethics; the logical positivists thought statements in ethics could not be verified and are, therefore, nonsense, nothing other than the statement of feelings. Popper disagreed; statements in metaphysics and philosophy, including those in ethics, are not scientific theories but they are nonetheless open to critical appraisal. Furthermore, the stand taken on philosophical issues has practical and political implications. For example, Popper's critical rationalism implies that in order to function well, science and societies need to facilitate criticism; this requires an open democratic society (Popper 2002). Popper is often identified with right-wing political movements and as he grew older he certainly became conservative. However, one of his friends and advocates, Bryan Magee, suggests that the open and critical approach he proposes "gives rise naturally to a radical attitude towards institutions" (Magee 1978, 266). Magee goes on to say that Popper shows radicals the danger of utopian thinking. The flaw in such thinking has two elements. The first is that if something is rotten it implies the whole system is rotten; for example, any system that allows people to starve when there is enough food is rotten. The second is that there must be something much better that can be put in its place. This leads to starting with an ideal and trying to get the world to fit it. Popper's political method, like his scientific method, is to begin from the problems you have and try to solve them. You will end up with new problems but you can make progress.

The philosophers I have mentioned with approval make uneasy bedfellows: Aristotle, Popper and, to a lesser extent, Bentham and Mill. Popper is critical of the essentialism he discerns in Aristotle; Mill developed a theory of inductive logic whilst Popper rejected any notion of inductive logic. In tackling issues in philosophy, I usually come round to Aristotle at some point; in this sense I feel more Aristotelian than anything else. And despite their differences there seem to be areas of similarity, for example, scientific realism. Other aspects seem to me compatible; Aristotle's method in philosophy seems to complement Popper's scientific method. Mill's ethics seems compatible with Aristotle's and, also, not to rely on the induction that Popper rejects. The most influential modern Aristotelian work for me so far has been Irwin's *Aristotle's First Principles* (Irwin 1990). This is a long, difficult book that I was able to read only because I was doing my doctorate. But the effort was worth it in giving me an in-the-round picture of Aristotle's philosophy and in starting to build my own. I seek to contribute to the philosophy of nursing by conveying my enthusiasm for the subject and for the writers that so impress me; perhaps also by conveying a belief that the subject matters. I suspect, though, that philosophy and the philosophy of nursing contribute more to me than I do to it.

5. Where do you see the field of philosophy of nursing to be headed, including the prospects for progress regarding the issues you take to be most important?

Autistic children can sometimes be identified in a group by their tendency to parallel play. They are in the group but not of it, often doing something repetitive and unconnected to the others. Autistic adults might talk at length on their favoured topic but if interrupted are either puzzled or simply carry on. At the moment, writers in the philosophy of nursing seem to be engaged in parallel play. There are two playgrounds: one contains those sympathetic to continental philosophy, the second, those closer to the Anglo-American tradition. Some in one playground might occasionally run over to the other to give them a fright. But even in the same playgrounds we seem to be pursuing our own lines of thought rather than building an area of inquiry. I have little sense of clear lines of argument and ideas being developed. At the moment the articles in *Nursing Philosophy* are like fireworks, exploding briefly to greater or lesser effect, but then disappearing. I would like to see more coherent bodies of work developing although I suspect in order to do this we would be better to talk of *Health Care Philosophy* or some such. I also wonder whether the work we do in philosophy could have some influence on practice. We might already affect education, in that students are sometimes directed to our articles; we might also affect research at least insofar as post-graduate nurses seem, perhaps are, compelled to write a chapter on methodology in which philosophy of science plays a big role. But I can't think of examples where nursing practice has been much influenced by work in philosophy, nor of a demand from practitioners that a particular philosophical issue be investigated. I would like this to be otherwise; for example, I would like my ideas on public health to influence nurses in the stand they take on individual blame for illness (Allmark and Tod 2007). I would like John Paley's ideas on phenomenology to influence researchers' choice of method (Paley 2005). For this to happen, perhaps one area for future work is the investigation of what, in other contexts, has been called knowledge-into-practice.

References

Allmark, P. 1995. "Can There be an Ethics of Care?" *Journal of Medical Ethics* 21(1): 19-24.

Allmark, P, and R. Klarzynski. 1992. "The Case Against Nurse Advocacy." *British Journal of Nursing* 2(1): 33-36.

Allmark, P, and A. Tod. 2007. Philosophy and Health Education: The Case of Lung Cancer and Smoking. In *The Philosophy of Nurse Education* edited by J. S. Drummond and P. Standish, 46-58. Houndmills, Basingstoke: Palgrave Macmillan.

Crisp, R. 2010. "Rights, Happiness and God: A Response to Justice: Rights and Wrongs." *Studies in Christian Ethics* 23: 156-163.

Le Fanu, J. 2011. *The Rise & Fall of Modern Medicine*. 2nd ed. London: Abacus.

Irwin, T. 1990. *Aristotle's First Principles*. Oxford: Oxford University Press.

Lynch, J. W., G. D. Smith, G. A. Kaplan, and J. S. House. 2000. "Income Inequality and Mortality: Importance to Health of Individual Income, Psychosocial Environment, or Material Conditions." *BMJ* 320: 1200-4.

Mackenbach, Johan P. 2012. "The Persistence of Health Inequalities in Modern Welfare States: The Explanation of a Paradox." *Social Science & Medicine* 75: 761-769.

Magee, B, ed. 1978. *Men of Ideas*. London: BBC Books.

Marmot, M. 2010. *Fair Society, Healthy Lives: Strategic Review of Health Inequalities in England post 2010*. London: Department of Health. Accessed on July 10, 2012 www. marmotreview. org

Nussbaum, M. 2006. *Frontiers of Justice:* USA: Harvard University Press.

Nussbaum, M, and A Sen. 1993. *The Quality of Life*. Oxford: Clarendon.

Paley, J. 2005. "Phenomenology as Rhetoric." *Nursing Inquiry* 12 (2): 106- 116.

Pope, C. n. d. "The Mischievous Ambiguity of the Word 'Poor'": Classification and Control in Oliver Twist and the New Poor Law.
Accessed on July 10th, 2012:
http://www. catherinepope. co. uk/downloads/Core Essay One. pdf.

Popper, K. 2002. *The Open Society and its Enemies: Volume 1: The Spell of Plato*. London: Routledge.

Quine, W. 1960. *Word and Object*. Massachusetts: MIT Press.

Wilkinson, R, and K Pickett. 2008. *The Spirit Level: Why More Equal Societies Almost Always Do Better*. Harmondsworth: Allen Lane/The Penguin Press.

3

Patricia Benner

Professor Emeritus

School of Nursing, University of California San Francisco, California, USA

1. How were you initially drawn to philosophical issues regarding nursing?

My earliest encounter with philosophy came during post-baccalaureate and Master's in Nursing education while studying cultural anthropology where different theories of meaning and language held my interest. I was particularly interested in Heidegger's notion of the links between language and dwelling in a culture.

In my doctoral studies, I began working with Professor Richard S. Lazarus on a project studying stress and coping in aging, which was for Richard Lazarus a move from the laboratory to field research. In this study of a community of relatively healthy, home-abiding middle aged to older adults, we kept bumping into the difficulties of studying mind-body-world links to stress, coping and health and moral outcomes. Dr. Lazarus, a truly wonderful mentor, and I had a very productive "misunderstanding" of his transactional model of stress and coping (Benner and Wrubel 1989) for three years. Dr. Lazarus' (Folkman and Lazarus1984) cognitive-phenomenological view of stress appraisals and coping was the most advanced and frequently used theory of stress and coping at the time. However, Lazarus held a more cognitive, Husserlian view of the mind-body-world relationship, a view that stayed within the Cartesian tradition and a rational-empirical approach to science.

I wrote a paper examining the impact of Richard Lazarus' influential and highly controversial early experiment that demonstrated that people had tacit knowledge of paired shock non-sense syllabi as indicated by galvanic skin responses to shock associated nonsense syllabi. Subjects perceived non-sense syllabi associated with shock as reflected by an elevated galvanic skin response. Non-sense syllabi associated with shocks, triggered a galvanic skin response in subjects receiving the mild shocks, even though the subjects could make no verbal, conceptual recognition of the syllabi associated with the mild shocks. This experiment led Lazarus to view primary stress *appraisals* as tacit, though he did not give up on the notion that they were *cognitive* mind based appraisals. At the time, I made the philosophical error of thinking that surely he must have thought that *tacit* just meant perceptual, embodied responses. This began our productive three year period of "misunderstanding" that led me to take "mind-body" philosophy courses with the noted mind-body-world philosopher, Professor Hubert L. Dreyfus (Dreyfus 1991) in order to think more clearly about *tacit awareness and* knowledge (Polanyi 1958).

That philosophical search was a major paradigm shift for me, changing my view of human sciences, mind-body-world relationships, the nature of perceptual grasp, embodied skilled know-how, the nature of dwelling in the lifeworld, stress and coping, and meanings, concerns, habits, and practices in everyday life. Phenomenological philosophy also led me to examine "high-end" coping of expertise based upon experiential learning from active practice in open-ended complex situations. This was all thanks to the articulate, inspiring philosophical teaching of Professor Hubert L. Dreyfus. I was also coached, and taught by Dr. Jane Rubin (1996) who was then his graduate teaching assistant and doctoral student. In each class on Kierkegaard, Merleau-Ponty and Heidegger, we delved into the difficult texts and proceeded under the tutorship of Professor Dreyfus to push out the boundaries of our understanding of these provocative, existential-phenomenological texts on mind-body-world relationships.

Meanwhile, I was steeped in the transactional, cognitive-phenomenological research, teaching and thinking of my primary mentor Professor Richard S. Lazarus. His theory of stress and coping, in its most condensed version, holds that primary appraisals consisted of the subject's *judgment* that the situation was one of harm, threat or challenge, i. e. stressful (benign *appraisals* were not considered stressful) (Folkman and Lazarus 1984). Secondary *appraisals* followed quickly and tacitly, had to do with how the person felt they could manage or weather the stressful situation. If the person immediately thought something to the effect: "I can handle this...I have managed much trickier or more difficult challenges in the past," then the primary stress appraisal response was ameliorated or even resolved to a non-threat or non-challenge pre-event stage.

In 1980 I wrote a chapter with Drs. Ethel Roskies and Richard Lazarus entitled "Stress and coping under the extreme conditions" (Benner, Roskies and Lazarus 1980). This chapter, focused on the experiences of the survivors of the Nazi Holocaust, transformed my understanding of mind-body-world relations and the links between the 'subjective' mind-body-world split of Cartesianism, and gave me a more social, phenomenological view. But without the philosophical teachings of Hubert L. Dreyfus and Jane Rubin, I would never have been able to articulate this shift in my understanding in intelligible language, or through the appropriate philosophical schools of thought related to the mind-body-world connections.

The phenomenological part of the Lazarus theory was Husserlian, and still lay within a Cartesian view of the mind-body-world relationship. This was central to the productive "mis-understanding" that I experienced related to Lazarus' theory. I was thinking in terms of Kierkegaard (2011), Heidegger (1962) and Merleau-Ponty's (1962) existential phenomenology of mind-body-world relationships. Stress appraisals were not conceptual or pure intentionality, rather they were quasi-emotive and embodied, a more Heideggerian and Merleau-Ponty version of tacit understanding and/or perceptual grasp. These two thought projects, one within coping and health and illness, and one studying expertise in any skilled know-how domain, have been the philosophical focus of all my work subsequent to this philosophical turn during my doctoral studies at University of California, Berkeley.

2. What, in your view, are the most interesting, important, or pressing problems in contemporary philosophy of nursing?

The following six issues are the most pressing problems in contemporary philosophy of nursing from my perspective:

1) *Addressing and augmenting and replacing the normalizing clinical gaze.*

Nursing has a non-normalizing and non-pathologizing tradition in practice, as evidence in the need to make "positive" diagnoses of strengths in the nursing diagnosis movement. Dr. Laurie Gottlieb has just published a landmark book entitled *Strengths Based Nursing, Health and Healing for Person and the Family* (Gottlieb 2012) that brings forth the Nightingale tradition of putting the body in the best conditions for healing. The clinical gaze and the normalizing approach is a psychologizing and deficit oriented approach as explicated by Foucault (1973) focused on diagnosing deficits rather than identifying positive and situated capacities and skills. This approach is pervasive in the medical/disease model of health care. Nursing as a practice, while strongly influenced by the clinical gaze and normalizing approach, has resistance to it in terms of helping people through the processes of recovery and re-entering and/or rebuilding their lifeworld after an episode of disruptive illness or injury.

In terms of health promotion, preventing illness and injury, and managing multiple chronic illnesses in aging societies, nursing has a strong alternative to a strictly illness oriented, tertiary care emphasis in health care. A better articulation of the philosophical assumptions that have driven a tertiary health for profit health care system could help with shifting the focus in health care making it more equitable, accessible and preventive. This a central philosophical issue in understanding and transforming our health care system into a more public health, preventive health care system. Currently medical education focuses on teaching and learning how to make complex differential diagnoses as the pinnacle of medical intellectual effort, despite the need for more focus on chronic illness management and preventive and public health care (Benner et al. 2009)

2) *Teaching and learning in nursing and other health care disciplines that encompass both the medical model and the lifeworld model of self-care and health promotion.*

A more sophisticated understanding of philosophy of science is needed that shows how the Cartesian allopathic model of illness care systematically excludes the social and physical environments of the person's world and the effects of the socially sentient, skillful, embodied person on health, illness and well-being (Benner 2000b; Gordon 1988; Sunvisson et al. 2009). Social medicine, medical anthropology and nursing have systematically tried to integrate the social with the medical, but with questionable success. A more sophisticated philosophy of science would help identify the significant distinctions between the social-human sciences and the physical sciences, and the implausibility of a single paradigm science in the social-human sciences. Human beings are finite and perspectival in terms of their historical situatedness and their ability to take a "God's eye view," and still engage in the study of daily engaged human lives of dwelling in a lifeworld. In the health and illness care of human beings, multiple perspectives are needed. A singular totalizing approach must be avoided for the sake of situated, contextualized and embodied understandings (Gallagher 2009). This is a central philosophy of science issue for nursing and medicine. Nursing, as a tradition and practice, has socially embedded knowledge and skills related to coaching persons through their recovery processes in relationship to the disease, injury, treatment demands, but also in relation to the person's lifeworld concerns and goals for their own being and functional capacities in the world.

3) *Nursing education: Integrating knowledge-acquisition and knowledge use and understanding clinical reasoning as a form of practical reasoning and understanding this reasoning as a source of knowledge in its own right.*

Nursing education has been hampered by a narrow rational-technical understanding of the practice of nursing, and also a focus on the use of formal and linear decision-making models of clinical reasoning. A narrow rational-technical view of thinking and knowledge causes nursing educators to pass over the productive, situated use of knowledge in practice (Gallagher 2009). Nurse educators, like most academics, gloss over the Aristotelian difference between *techne* and *phronesis*. *Techne* refers to what can be standardized and applied as a technique or procedure (Bourdieu 1992; Dunne 1997). *Phronesis* refers to wisdom and practical, situated reasoning: "Reasoning across time about the particular through changes in the person/situation and/or changes in the situation" (Benner, Hooper-Kyriakidis and Stannard 2011). Much of nursing practice is taught as if knowledge and knowing are a mere application of *applying* or using a technical-theoretical template on open-ended, under-determined clinical situations.

The major agenda in nursing education has been to generate nursing theories that can then be *applied* to nursing practice. This narrow rational-technical approach overlooks the ways in which nursing is a complex, under-determined practice with notions of good internal to that self-improving practice (MacIntyre 1981). Classroom and clinical education have been radically separated so that practice is, more often than not, viewed as a place of mere application of theory, science and technology learned in the class room, which is considered to be a place of didactic theoretical learning.

Another major agenda in academia (Sullivan and Rosin 2008) is an emphasis on critical thinking. There is a conflation in nursing education between diverse modes of problem solving and clinical reasoning and *critical thinking*. This is a misguided emphasis because most often in practice, nurses are required to engage primarily in clinical reasoning, and only in times of complete practice breakdown, novel problems or confusion are nurses required to engage in critical reasoning.

As noted earlier, clinical reasoning is a form of practical reasoning, i. e. reasoning across time about the particular through changes in the patient, and/or changes in the nurse's understanding of the patient situation. Practical reasoning is a form of reasoning through transitions (Benner 1994; Taylor 1993). Practical or clinical reasoning has been relegated to the margins in all of academia, not just nursing. The academic preference is for formal models of decision making and critical, deconstructive thinking.

All forms of reasoning are required in actual nursing practice - imaginative, creative thinking, clinical reasoning, critical thinking, scientific problem solving process, situated use of scientific evidence, and clinical reasoning - as nurses judge changes in the patient's progression or regression of illness, and titrate treatments and medications based upon the patient's responses to treatments, medications and to their illness. In everyday practice, nurses frequently discuss patients' trends and trajectories. Static, snap shot reasoning, as in scientific experiments, while useful for science, is neither practical nor suitable for actual nursing practice, which is always about the particular in relation to the general.

4) *A philosophical re-interpretation of the nature of human beings as active learners, and the importance of socialization and formation in nursing education.*

Understanding human beings, as socially constituted with shared meanings, and constitutive and prior to individual self-understandings and meanings, has profound implications for nursing education (Benner et al. 2009; Benner, Hooper-Kyriakidis and Stannard 2011). For example, if we accept a "socialization" view of education, we unwittingly leave out the role of the student as an agent seeking to contribute to his or her own self-understanding. Formation requires that we think of the learner's agency in taking up a practice. I have drawn on Margaret Mohrmann's (2006) metaphor to capture the embodied situated agency of formation in a practice discipline:

In the tradition of technical professionalism, professional educators use terms like socialization, role playing, and role taking to discuss the identity, character, and skill development of the professional. It is here that Educating Clergy, the first of the five Carnegie Foundation for the Advancement of Teaching studies to be published, focuses on how professionals are educated, providing a much needed and richer language, particularly so for education of nurses and physicians. Socialization into a set of norms, values, and styles of interaction and comportment cannot account for the constitutive and world-transforming nature of learning the skills, habits and practices of a profession. It does not capture the richness or complexity of learning ethical comportment and ways of being and understanding oneself as a nurse or physician. (Benner 2011, 6)

This view of formation is congruent with a civic professionalism view of the nature of professions.

5) *A better philosophical understanding and articulation of the nature of caring practices and their influence on growth and development, health, illness, and injury.*

From the beginning of my philosophical work (Benner 1982; Benner and Wrubel 1989), my goal has been to bring the articulation of caring practices in from the margins, as well to develop a non-trivializing, non-sentimentalizing and non-moralizing understanding of caring practices into mainstream nursing and public discourse. By caring practices, I mean the caregiving, nurturing, parenting, sponsoring work of supporting and caring for those who are dependent and require caregiving and nurturing, concerned with the ill, injured, the very young and very old, and all persons going through major developmental and event related changes, including everyday health and morale support that most people rely on in their relational communities.

I still consider this a pressing need both at the cultural levels and at the nursing discourse levels. As Nancy Sherman (1997) and Onora O'Neil (1996) point out, a good society requires justice and care. Those existing in highly individualistic western cultures have a tendency to pathologize the need for care, and caregiving itself. While there are always many potential pathologies of helping, such as trying to dominate, control, or foster dependency, these problems make the need to better articulate and support healthy and effective caring practices even more important (Benner 2000a, 2001; Sunvisson et al. 2009). A better public articulation and understanding of the primacy of caring and caring practices are essential to better, richer understandings of the nature of dwelling in lifeworlds.

Articulating a richer more robust account of embodied skilled knowhow, emotional reasoning and the influence of habits, skills and practices on coping, human flourishing and dwelling in self-improving, growth-oriented, safe and sustainable lifeworlds, is central to the project of understanding and fostering caring practices and a healthier, more just and human society. This is a phenomenological understanding of what it means to be human, dwelling in lifeworlds suspended by relationships, concerns, and everyday dwelling. Human lifeworlds are systematically excluded from allopathic, Cartesian medicine. Bringing our understanding of caring practices in from the margins of nursing and public discourse can also update, clarify and enrich our understanding the central roles of lifeworld, the social, sentient lived body, life story, and social identity in growth and development across the lifespan and coping and recovering from an illness or injury.

6) *Inter-professional practice, civic professionalism, and better understanding of clinical and scientific ways of understanding and the problems associated with commodification in market-based health care systems*. [see response to question 3 below]

3. What, if any, practical and/or socio-political obligations follow from studying nursing from a philosophical perspective?

As many philosophers have noted, knowledge is power, and as Foucault pointed out, in our historical era we are caught in everyday micro-processes of power, self-making, and self-control. Nursing is rooted in the power relationships of gendered roles, and in an extremely hierarchical medical industrial complex. Nurses are front-line health care providers and care for persons across the spectrum of health and illness. It will not help just to reverse the oppositional powers and status of nursing and medicine, anymore than it will help to do so in any oppositional system where one side is defined in terms of the other so that the subordinate "side" is defined in relation to the dominant side, instead of being defined in its own terms. Defining nursing in terms of practical wisdom and skilled know-how, and articulating knowledge and skills and wisdom embedded in everyday expert nursing, i. e. in relation to the content, intents and process of actual nursing practice, could create a new definition and understanding continuum of knowledge, care, skills, habits and practices among the professional health care team. Understanding all health care practices in terms of civic professionalism instead of the competitive individualism and technicism of 'technical professionalism' could transform the social injustices in health care access, and health care outcomes between the poor and marginalized minorities. Understanding health care practices in terms of civic professionalism (Sullivan 2004) could infuse new notions of good into the practice disciplines, so that they better understand themselves, not in terms of power, competitive individualism and capitalism, but rather in terms of their civic responsibilities to the society. A democracy depends on the civic responsibilities of the professional classes, but the civic responsibilities of the professionals become marginalized in a system that focuses on the technical powers and science of the practitioners and minimizes their civic (citizen) responsibilities for creating a just and good society.

Such a philosophy demands a more liberal, communitarian view of society, and a more active, generous participant membership of citizens, instead of the view of society and health care as competitive purveyors and consumers of a health care industrial complex. Health care, when commodified, turns patients and citizens into consumers. Yet the need of care during dependency, for example in childhood or old age, illness and injury, strips the person of their 'consumer powers' and they willingly or not become supplicants, and reduced to the economic and power structures of the health care industrial complex. Health care defined in terms of illness 'product lines', such as 'cancer product lines' and 'heart surgery product lines', truly forgets the vulnerabilities and resistances of those who must submit, however reluctantly, to the use of those unwanted, but coercive and controlling powers. The ill or injured experience themselves in their most vulnerable times to be, in differing degrees, at the mercy of the 'sellers' of the best product lines. Hopefully they are fully protected from economic ruin through purchased insurance coverage through at least one of some 5,000 providers.

4. In what ways does your work seek to contribute to philosophy of nursing?

I do not believe that the above issues that I consider essential to the progress of nursing science and practice have yet been fully realized. Inroads have been made however. Caring practices are increasingly being examined in nursing and the social sciences, though few intellectuals still hold to the notion of a computer model of the mind and skilled know-how. Instead there is now, notably, a whole *Cambridge Handbook on Situated Cognition* and embodied intelligence. The discourse in the social sciences is changing more quickly than the narrow rational-technical discourse that continues to be in ascendancy in nursing science, practice and education. Nursing has much to contribute to philosophy, just as philosophy has much to contribute to nursing. The actual practice of nursing with its deep understandings of illness, injury and recovery has much to offer philosophy in terms of mind-body-world relationships, well-being, and capacities for dwelling in lifeworlds.

I believe that my greatest contribution thus far in philosophy of nursing has been to raise a new level of consciousness about the mind-body-world relationships that influence health and illness. Perhaps the largest impact of my work has been in the ongoing development of the notion of nursing practice as a way of knowing in its own right and the importance of articulating what we know in our practice, for the enrichment and heightened relevance of our nursing research questions and programs. Thus far my work has had an impact in developing knowledge and clinical nursing inquiry in nursing practice (Benner 2000b; Benner, Tanner and Chesla 2010; Benner, Hooper-Kyriakidis and Stannard 2011), in bringing the relationship between knowledge acquisition and knowledge use (a form of productive thinking) together in nursing education, and in articulating transformative pedagogical strategies that are specific to nursing in the recent Carnegie Foundation for the Advancement of Teaching publication (Benner, Sutphen, Leonard, and Day 2009). These nursing domain specific pedagogies include teaching for a sense of salience, clinical reasoning as a form of practical reasoning as central to nursing practice, the need for multiple frames of reference, multiple problem solving approaches with an emphasis on clinical reasoning, and the role of formation and situated coaching as essential to nursing education.

5. Where do you see the field of philosophy of nursing to be headed, including the prospects for progress regarding the issues you take to be most important?

A more fully articulated and original philosophy of nursing is in its relative infancy. Published works in nursing philosophy have been hampered by lack of training at the doctoral level by many nurses authoring philosophical positions intended to be all inclusive and totalizing. Critiques from a philosophical stance are often made without seeking to accurately understand and articulate the philosophical position being critiqued. The voice of nurse writers of philosophy has sometimes been strident and lacking questioning and curiosity. These are the problems that plague philosophical understanding of any practice discipline that actively seeks to understand themselves more in terms of a science, rather than a philosophy. Philosophical assumptions, immersed in rational empiricism, are overlooked. Questions of the social, sentient, quasi-emotional nature of situated thinking in action and practical reasoning are overlooked in such a narrow scientistic view. Equally troubling are untenable, pseudo-scientific understandings, posed philosophically without providing means of falsification or justification.

However, I am convinced that meta-theoretical thinking is essential to the progress in creating and evaluating nursing knowledge and practice. We need to understand the root metaphors we are using to understand health, illness, persons, physical environments, skilled know-how, concerns, practices and lifeworlds. The practice of nursing is diverse, broad, and rich in understandings of stress, suffering, and coping, along with recovery possibilities, and the situated possibilities of anyone recovering from an illness or injury. We need to move beyond a narrow allopathic Cartesian model of medicine, nursing and health care. Rigorous, challenging, curious philosophical approaches can help us do that. Articulating what we come to know in our practice, and understanding that our practice is a way of knowing in its own terms, can bring new and old philosophical inquires into view. We have not adequately studied embodied dwelling in the world in cumulative and systematic ways. Narrative self-understandings and human concerns and challenges in recovering from illness and injury have been poorly articulated from the perspective of the one experiencing the recovery of self and world. I fervently believe that philosophy of nursing can bring marginalized caring practices and marginalized understandings of dwelling in a lifeworld into view in ways that could transform our health care system.

References

Benner, P. 1982. "From Novice to Expert." *American Journal of Nursing*, 82: 402-407.

Benner, P. 1994. "The Role of Articulation in Understanding Practice and Experience as Sources of Knowledge." In *Philosophy in a Time of Pluralism: Perspectives on the Philosophy of Charles Taylor* edited by J. Tully and D. M. Weinstock. Cambridge: Cambridge University Press.

Benner, P. 2000a. "The Quest for Control and the Possibilities of Care." In *Heidegger, Coping and Cognitive Science: Essays in Honor of Hubert L. Dreyfus, Vol. 2* edited by M. Wrathall and J. Malpas. Cambridge: MIT Press.

Benner, P. 2000b. "The Roles of Embodiment, Emotion and Lifeworld for Rationality and Agency in Nursing Practice." *Nursing Philosophy* 1(1): 5-19.

Benner, P. 2001. "The phenomenon of care." In *Handbook of Phenomenology and Medicine, Vol 68*, Philosophy and Medicine Series edited by S. K. Toombs, 351-369. Dordrecht, The Netherlands: Kluwer Academic Publishers.

Benner, P. 2011. "Formation in Professional Education: An Examination of the Relationship between Theories of Meaning and Theories of the Self." *Journal of Medical Philosophy* 36(4): 342-53.

Benner, P., Hooper-Kyriakidis, P. and D. Stannard. 2011. *Clinical Wisdom and Interventions in Acute and Critical Care: A Thinking-in-action Approach*. New York: Springer.

Benner, P., S. Janson-Bjerklie, S. Ferketich, and G. Becker, G. 1994. "Moral Dimensions of Living with a Chronic Illness, Autonomy, Responsibility and the Limits of Control." In *Interpretive Phenomenology: Embodiment, Caring and Ethics* edited by P. Benner, 225-54. Thousand Oaks, CA: Sage.

Benner, P., E. Roskies and R. Lazarus. 1980 "Stress and Coping Under Extreme Conditions." In *Survivors, Victims and Perpetrators: Essays on the Nazi Holocaust* edited by J. E. Dimsdale, 219-258. New York: Hemisphere Publishing Company.

Benner, P., M. Sutphen, V. Leonard, and L. Day. 2009. *Educating Nurses: A Call for Radical Transformation*. San Francisco: Jossey-Bass and Carnegie Foundation for the Advancement of Teaching.

Benner, P., Tanner, C. A. and C. A. Chesla. 2009. *Expertise in Nursing Practice: Caring, Clinical Judgment and Ethics*. New York: Springer.

Benner, P. and J. Wrubel. 1982. "Skilled Clinical Knowledge: The Value of Perceptual Awareness." *Nursing Educator* 7(3): 11-17.

Benner, P. and J. Wrubel. 1989. *The Primacy of Caring: Stress and Coping in Health and Illness*. Upper Saddleback, NJ: Prentice-Hall

Bourdieu, P. 1992. *The Logic of Practice,* translated by R. Nice. Stanford, CA: Stanford University.

Dreyfus, H. L. 1991. *Being and Time: A Commentary on Division I.* Boston, MA: M. I. T. University Press.

Dreyfus, H. L., S. E. Dreyfus, and P. Benner. 2009. "Implications of the Phenomenology of Expertise for Teaching and Learning Everyday Skillful Ethical Comportment." In *Expertise in Nursing Practice, Caring, Clinical Judgment and Ethics* edited by P. Benner, C. Tanner, and C. Chesla, 258-279. New York: Springer.

Dunne, J. 1997. *Back to the Rough Ground: Practical Judgement and the Lure of Technique*. Notre Dame: University of Notre Dame Press.

Folkman, S. and R. S. Lazarus. 1984. *Stress, Appraisal, and Coping*. New York: Springer

Foucault, M. 1973. *Birth of the Clinic. An Archaeology of Medical Perception*. New York: Vintage Press.

Gallagher, S. 2009. "Philosophical Antecedents of Situated Cognition." In *The Cambridge Handbook of Situated Cognition* edited by P. Robbins and A. Murat. Cambridge: Cambridge University Press.

Gordon, D. 1988. "Tenacious Assumptions in Western Medicine." In *Biomedicine Re-examined* edited by M. Lock and D. Gordon, 19-56. Boston: Kluwer.

Gottlieb, L. 2012. *Strengths Based Nursing, Health and Healing for Person and the Family.* New York: Springer

Heidegger, M. S. 1962. *Being and Time,* translated by J. Macquarrie and E. Robinson. New York: Harper and Row.

Kierkegaard, S. 2011. *Fear and Trembling*. New York: Penguin Classics.

MacIntyre, A. 1981. *After Virtue: A Study in Moral Theory.* Notre Dame, IN: University of Notre Dame.

Mohrmann, M. E. 2006. "On Being True to Form." In *Health and Human Flourishing: Religion, Medicine, and Moral Anthropology*, edited by C. Taylor and R. Dell'Oro, pp. 9-102. Washington D. C.: Georgetown Univ. Press.

Merleau-Ponty, M. 1962. *The Phenomenology of Perception* translated by C. Smith. London: Routledge & Kegan Paul.

O'Neill, O. 1996. *Towards Justice and Virtue: A Constructive Account of Practical Reasoning*. Cambridge: UK: Cambridge University Press.

Polanyi, M. 1958. *Personal Knowledge: Towards a Post-Critical Philosophy*. London: Routledge & Kegan Paul.

Rosch, E. 1983. "Prototype Classification and Logical Classification: The Two Systems." In *New Trends in Cognitive Representation: Challenges to Piaget's Theory* edited by E. Schnolnick, 73-96. Hillsdale. N. J.: Erlbaum.

Rubin, J. 1996. "Impediments to the Development of Clinical Knowledge and Ethical Judgment in Critical Care Nursing. In *Expertise in Nursing Practice, Caring, Clinical Judgment and Ethics* edited by P. Benner, C. Tanner, and C. Chesla, 170-192. New York: Springer.

Sherman, N. 1997. *Making a Necessity of Virtue: Aristotle and Kant on Virtue*. Cambridge, UK: Cambridge University Press.

Sullivan, W. M. 2004. *Work and Integrity: The Crisis and Promise of Professionalism in America*. San Francisco: Jossey-Bass.

Sullivan, W. and M. Rosin. 2008. *A New Agenda for Higher Education: Shaping a Life of the Mind for Practice*. San Francisco: Jossey-Bass.

Sunvisson, H., B. Habermann, S. Weiss, and P. Benner. 2009. "Augmenting the Cartesian Medical Discourse with an Understanding of the Person's Lifeworld, Lived body, Life Story and Social Identity." *Nursing Philosophy* 10(4): 241-53.

Taylor, C. 1985. "Theories of Meaning." In *Human Agency and Language. Philosophical Papers, Vol. 1* edited by C. Taylor, 248-292. Cambridge: Cambridge University Press

Taylor, C. 1993. "Explanation and Practical Reason." In *The Quality of Life* edited by M. Nussbaum and A. Sen, 208-231. Oxford: Clarendon.

4

Karin Dahlberg

Professor of Caring Science

Linnaeus University, School of Health and Caring Sciences, Lifeworld Centre for Health, Care and Learning

University of Gothenburg, Centre for Person-Centred Care, Sweden

---

1. How were you initially drawn to philosophical issues regarding nursing?

I have always been drawn to philosophical issues, always been wondering about "the world" and what it means; what humans, animals and plants are, and what the differences are. I have always asked questions, such as 'why is it so'? 'How does it work'? More questions than parents and teachers generally liked.

When I was a nursing student in the 1970s, I asked questions about the philosophy of medicine and the philosophy of illnesses but got no satisfying answers. I was met by responses such as: "This is as it is. You only need to learn this method, or these principles." Nursing education then was a technocratic business, added by natural science facts on the one hand, and behaviorist theories and methods on the other.

I chose psychiatry as my specialty and my clinical work was as a mental health nurse. I met patients as well as professional carers, who understood the world and mental illness so differently compared to anything I had previously experienced. I was stunned by the patients' narratives and the various interpretations of these narratives among the professionals. I asked questions of all people everywhere all the time but had to consult the literature in order to really understand. I headed towards psychotherapy and began reading into psychology and therapy. I wondered how the patients' inner worlds were built up and how they may have differed from mine, how our worlds could communicate and how they could best be supported. These last questions have remained the focus for me to this day.

After a few years as a nurse I still had far more questions than answers. I elected to enter school again and chose the health care teacher education route. At the department where I was a student-teacher, the subject of pedagogy had been an active one for at least a decade. There were researchers who rebelled against the modernist approach to teaching and learning, which they saw as didactic and technorational. Under the leadership of Professor Ference Marton they developed what they called Phenomenography, a West-Swedish enterprise implicitly inspired by the phenomenological movement. I still clearly remember the very day when one person from the Phenomenography group came to us young student-teachers, and asked if we knew what knowledge and learning is. After listening to our ponderings upon this for a while, they stated that "learning" simply means "to look differently upon something in the world around us." We were astonished, almost startled. Several students were deeply concerned and uneasy by such an apparently unusual way of understanding pedagogy. Only a few of us became interested and I was the only one who was seriously drawn to this idea.

As a consequence, I gave up the psychotherapy path and went for pedagogy, where I found theories close to psychology and psychotherapy that gave me new perspectives on the human mind, thinking and consciousness. In the middle of the 1980s I was accepted as PhD student in Pedagogy and joined the Phenomenography group. Even if the PhD education and thesis writing mostly meant empirical research, what really thrilled me was the philosophy that I encountered, and in particular epistemology. New questions arose: What is knowledge? Are there different sorts of knowledge? Is there knowledge that is better, e. g. more scientific than other forms of knowledge when we want to understand human worlds? How is it possible to obtain knowledge about other minds? Are there different ways to find different forms of knowledge?

I had earlier attended courses in philosophy of science and was acquainted with hermeneutics. However, I needed more insights into epistemology and I was not contented until I began to study phenomenology. It was at this moment, for me, that all the world view horizons became clearer. I had many new questions but, finally, I also got some answers.

In retrospect, I can see that all perspectives, theories and different kinds of texts that heavily addressed me have something in common: namely, and in general terms, what we call the continental approach, and in particular the philosophy of phenomenology. That is the common thread of my main interests, such as nursing philosophy in the form of the Caring Science approach, Human Science Research approaches, more of which below.

2. What, in your view, are the most interesting, important, or pressing problems in contemporary philosophy of nursing?

The retrospective account given above coheres with my present view that many of the old problems in health care and nursing still exist. We have seen changes and improvements but the interesting, important, or pressing problems in contemporary nursing and its philosophy are the same as the old ones, or are related to them.

In Europe, modern medicine was developed in congruence with the significant evolution of science. A mechanistic world view directed both medicine and science: man was described as a sophisticated machine in a mechanistic world, with basically simple laws of cause and effect, or at least such was the impression given. As with the natural sciences, medicine's celebrated successes were built up by new knowledge that helped develop new cures for previously incurable diseases. Drug companies and pharmacies grew and became lucrative. Nurses became assistants in this thriving business. Even if they saw other needs and had lived experiences of different care and treatments, they adapted to the ideals and work of modern medicine. This indeed was part and parcel of their progress into modernity.

In the early 1980s in Sweden, health care education, in particular nursing, physiotherapy and occupational therapy became part of the university system. At the same time there was an educative shift in favor of a more human sciences oriented view. People were no longer to be addressed in pieces and fragments but were rather to be seen as the complex living existences they are. However, the effect on nursing practice was limited. The most significant result was perhaps the many debates and publications which had the aim of identifying and overcoming the perceived gap between theory and practice.

An explanation must be sought for in both sides of the perceived gap. In the universities, theories and concepts have been developed in a way that arguably make sense in theory but are often perceived to be too far from the everyday reality of care (Ekebergh 2001; Ranheim 2011). In Sweden, the long time in academic isolation has arguably made this situation worse. Some theories and concepts seem far removed from the pragmatics of everyday care. The language seems unfamiliar, even strange to the practitioners. For nurses in particular, the tendency not to continuously read and learn in practice makes this issue more pressing.

Care is still dominated by medicine as a scientistic and biological paradigm. Quite simple cause and effect models are still in focus and are hard to adapt to for those nurses who want to walk a different road. Unfortunately, nurses are in many ways still assistants to the physicians, perhaps because they want to be, or even if they do not want to be.

However, we must add to this the impact of Evidence-Based Medicine (EBM), which is designed in such a way that it is hindering vital research and development in health care, not least in nursing. An example: Suffering, to be studied within the EBM model, must consist of one single measurable symptom that must be related to a standardized treatment intervention. This is clearly problematic in that people's emotional world is more complex and diverse and with a higher level of inter-related experiences than the assumptions that EBM describe. Sensations of, for example, pain, anxiety, sorrow and depression are often simultaneous and are experienced as "ill-being", "discomfort" or "malaise". Emotions can neither be reduced to one simple aspect nor simply measured. Further, care cannot and should not always be standardized. An important point within nursing is that care must adapt to the present individual in a current situation. This must be mirrored in nursing and caring science research.

One of several consequences of the EBM dominance is that we have large and increasing drug consumption. It is logical to measure the effect of a drug on the basis of EBM, and it is difficult, if not impossible, to highlight other treatments or such care methods that could replace or supplement drug treatment. The proven experience suggests for example that it is often possible to help a troubled patient to sleep with some small talk at the bedside, perhaps supplemented with a glass of milk or a foot bath with calming oils. It is, however, more difficult to isolate the effect of good nursing care than a pill with current evidence-logic. As a consequence, methods for caring such as massage, caring touch, music, movement or art therapy or existentially focused dialogues cannot compete with the medical models. The lived body simply does not fit the current EBM model and is, consequently, neglected. This is a huge problem for nursing, which could be so much more efficient and support people's health processes so much better.

The currently accepted EBM model has several weaknesses and serious problems and consequences, which are important to identify. A key element is that many researchers (and practitioners) in health care lack knowledge of scientific theory and methodology. In Sweden today, it is possible to study for a PhD without any study of philosophy of science. For comparison, consider how it would be if one had engineering training without a single point in mathematics or physics.

Nursing needs more and better scientific training. There are significant gaps in both "quantitative" and "qualitative" methods education. The statistical accuracy must be improved, the database and level of analysis must be calibrated with greater precision, and a discussion of how statistical results can be interpreted for application in practice must increase. The high prevalence of "content analysis" must be qualified by a more informed and in-depth account of qualitative approaches, and we need more well-informed discussions of what evidence means in caring science and nursing research, not least in relation to "qualitative methods". Perhaps philosophy should seek to play a credible and more considered role in clarifying this ground.

3. What, if any, practical and/or socio-political obligations follow from studying nursing from a philosophical perspective?

In philosophy of nursing, both researchers and practitioners find arguments for the development of care in theory as well as in practice, arguments that could and should be used. In philosophy we find ontology, epistemology and methodology and can develop insights in all nursing areas, e. g. on caring ethics, caring methods and research. In the current situation, when the new magic of numbers and the measurement frenzy is all over us, it is philosophy that can guide us to new and better understandings of what it means to exist in the human world, both considering what health in general demands from us and not least when being in illness. This at least would be a considered obligation.

4. In what ways does your work seek to contribute to philosophy of nursing?

To me it is an indefatigable truth that philosophy is a solid foundation for any theoretical or practical endeavor within any scientific field. Without the illumination of the ontology, epistemology and methodology of caring and nursing, or caring and nursing research, we are like reeds shaken by the wind. As nurses, teachers or researchers of health, illness and caring we all need such sound foundation, and tools for our work, that philosophy offers.

My contributions are mainly two, published as *Reflective Lifeworld Research* (2001, 2008) and *Hälsa och Vårdande i Teori och Praxis [Health and Caring in Theory and Practice]* (2010).

The first work is the result of my main focus on epistemology and methodology that originally developed due to the struggle to understand Phenomenography and its place in the world of science and research in general and qualitative methodology in particular. The implicit relationship to phenomenology was too subtle for me and I began my very long but wonderful journey through continental philosophy, with a focus on phenomenology.

With *Reflective Lifeworld Research* (2008), written together with my daughter Helena who is a philosopher and Merleau-Ponty expert, and Maria Nyström who actively practices the hermeneutic research approach, my aim was to outline and thoroughly describe a research approach that could support human science research and, not least, nursing research. The approach should serve researchers who want to better understand, describe and interpret health, illness and caring phenomena that are characterized by high levels of complexity, inter-relationships and richness of detail. The emphasis is on the Lifeworld, human intentionality and its capacity for seeing meaning as well as for reflection.

At the time of the first edition of *Reflective Lifeworld Research* (2001) I had the experience of advising PhD students and of observing how much they benefited from studies in philosophy of science in general and phenomenological epistemology in particular. Therefore I completed an analysis of the pertinent philosophy of Edmund Husserl, Maurice Merleau-Ponty, and Hans-Georg Gadamer in particular, but also some texts by Martin Heidegger. Essentially the book draws on the theories of lifeworld (including co-existence), intentionality (including the idea of co-constitution), and the theory of the lived body (based in the idea of the "flesh of the world").

The choice of philosophical theory and the description of the research approach were inspired by the experiences of editing and reviewing articles. In the 2001 edition of *Reflective Lifeworld Research*, Nancy Drew contributed a North American perspective and experiences from North American nursing. Our common experiences served us with insights into the fallacies that were and still are dominant in nursing literature. We reacted against the weakly described approaches, the mixed discourses and the shallow use of methodology. Of course, nothing is permanent. As time moved on, we knew that there was further work to be done. At times roughly expressed ideas in the first edition were ready for refinement and improvement.

This led to the second and revised edition (2008). The evidence debate had grown stronger and EBM was an unavoidable concept. As a consequence, the new edition included some of these discussions. However, by that time I had not myself completed a full analysis of the idea of evidence, a project that I am about to finish this year. I think it is of great importance that nurses are able to understand the idea of evidence and find ways of assuring evidence in their research, so that caring can develop and patients can avail themselves to more lifeworld sensitive caring methods that support health and well-being, even when illness is a fact. It is for this reason that we must allow for the concept of Evidence Based Practice (EBP) including nursing practice that goes beyond EBM to embrace Lifeworld activities and interventions that can be shown to make a difference.

It is in this respect, that the other main contribution that I want to present is a new theory of Health and Caring. The theory has two important foundations: The first one is the empirical experience from many research projects, of which an important source is the research done together with 17 PhD students. Little by little during the last ten years I realized that even if all the findings were published and all research was rewarded in that way, there was still more to do. The different research reports indicated findings beyond the single projects that focused upon many aspects of nursing. Through all the diverse contextual meanings of health and caring there was something more essential to health and caring to be illuminated.

Throughout the years of philosophical analysis with the aim of understanding science and research it became clear that the philosophical texts that I found not only bestow on us epistemology and methodology but also ontology. With the point of departure still in the rigorous analyses of Husserl, I wanted to go back to the early interests on the meaning of life and existence, and therefore deepened my reading of texts by Merleau-Ponty, Sartre and Heidegger and other philosophers in the area of existential theory. During one of my trips to the US giving summer courses on phenomenology, I found Hans-Georg Gadamer's (1996) little known book *The Enigma of Health: The Art of Healing in a Scientific Age*. I then understood that my growing insights in that direction were not completely wrong.

A few years ago I was excited enough about this big project to give my university notice of my departure. I ended the safe, well-paid, and very interesting employment as professor and director of the PhD program. The project was too big, too demanding and too necessary to be treated as a side-job; it needed to be the main duty for one year.

During the work with *Reflective Lifeworld Research* (RLR) I had also tried to understand different research approaches such as Ethnography and Grounded Theory, and a strategy that I assumed was quite possible within a phenomenological framework, but not one that I had experienced myself, was to ground a theory in several instances of empirical research. Now the time had come for this challenge of the RLR approach.

It meant hard but inspiring work to wade through all the empirical descriptions, all meanings of health and caring that were embedded in the observations and interviews with patients and their close relatives, as well as with carers and most of all nurses. The meta-analysis ended up in a new meaning structure of health and caring, which included sometimes severe illness and different nursing contexts. The empirical findings were then further reworked by the philosophy, which worked like a giant spotlight, illuminating all dark spots of the empirical description.

My new theory reveals health, as the basic motif of caring, as the experience of well-being and a quality of being able to. In order for health to be there must be a sense of well-being in life, which includes individuals' capacity to realize small as well as big life projects, together with their further involvements in the world of others. The aim of caring according to this theory is to support and strengthen people's health processes, but the empirical analysis, as well as Gadamer (1996), show that health is an enigmatic, taken-for-granted and un-reflected phenomenon that most commonly is ignored or vaguely described. Instead, many discussions and publications on health and caring are related to illness (see for example Peters et al 2007).

In this new theory, health is related to existence and the everyday life. The essential themes of health are described in terms of life power, life rhythm and life context. Here is also the place for notions on movement and stillness as well as safety and loneliness. All aspects of health are inter-related. For example, movement has always to be understood as figure against a background of stillness, and stillness has to be understood as a figure against a background of movement. As another example, loneliness has to be understood in relation to a sense of belongingness, with others or with "some-thing" in the world.

The last decade has shown a new emphasis on person-centred or patient-centred care. Even in the new theory the patient perspective is in focus as an ethical approach to health and caring. However, to not fall into the trap of viewing patients as "care customers" on the one side, or the paternalistic perspective on the other side, where patients are passive receivers of care, this theory has as a solid foundation in the philosophical idea of existence as lifeworld. The theory deals with how to access another's lifeworld, even in hard situations when caring is engaging patients without language, or when the lifeworld is very different from the commonplace understanding. To further confirm the existential depth of people's health processes the theory points to the freedom and vulnerability that characterise human life, and draws on existential ideas of definiteness and indefiniteness.[1]

My ambition with taking on this project was to bring together theory and practice. This is the point where the description of the theory became a book on health and caring and a second author was invited in the project. The variables of caring that support health, well-being and the quality of being able to are described by common philosophical concepts such as openness, intersubjective (inter-corporeal) relationships, caring encounters, caring communication and caring presence. Basic caring tools are dialogues in different contexts and caring treatments, e. g. caring touch, massage. The important everyday care must include life-rhythms, with healthy portions of movement and stillness. Here are given some examples of such existential and everyday ingredients as

[1] This part of the theory has partly been developed in caring science collaboration between The Bournemouth University in Great Britain and Linnaeus University in Sweden (cf. Todres, Dahlberg & Galvin, 2007; Dahlberg, Todres & Galvin, 2009).

walking and dancing as well as Asian ideas e. g. yoga. The majority nowadays are suffering from too little sleep and the experience of how it is hard to protect oneself against the bombastic attacks of sensory stimulation. Finally, nursing comes into the picture, together with occupational therapists, physicians and physiotherapists. Against the background of the theory, caring activities are presented in relation to the four different professions. One emphasis is on caring activities that are neglected by modernist medicine and health care.

5. Where do you see the field of philosophy of nursing to be headed, including the prospects for progress regarding the issues you take to be most important?

At present we face a quite unproductive fight between scholars who talk about "nursing" and "nursing science research" on the one side and those who fight for "caring" and "caring science" on the other. My approach can be described as neither of these in absolute terms. I use the concepts "caring" and "caring science" at the same time as I emphasize the importance of nursing. The rationale for using caring science concepts is two-folded.

For a long time I have participated in, read about and reflected over the rapid development of academic nursing. However, I have many times been sad about the close relationship to medicine, as if we have not been able to build our own house of science, research and practice. Too many publications have focused upon diseases, illness and diagnoses, and consequently the literature is filled with descriptions of what nurses should do or not do. This has not happened without debate and there are alternatives. Other scholars have presented ideas of "being" as an opposition to "doing". However, these ideas have often been accused of being characterized by "new age" dimensions and it has been argued that they need to come down to earth, to be more practical, more put into clinical use and evaluated (cf. Hallberg 2006). Hallberg (2006) focuses upon an important dilemma in nursing research and I agree that we need to work for more impact on clinical nursing, but I don't agree with her limited call for more quantitative research, which means, as I argue above, that many important nursing tools are being lost.

An article was published as a reply to Hallberg's guest editorial (see Galvin et al. 2008). Our comments were born from the work within the European Academy of Caring Science (EACS)[2] where we argue for the value of caring science that is not limited by professional domains but that focuses upon the patients and their essential participation in health and caring processes.

[2] EACS http://www2. pubcare. uu. se/care/eacs/

I claim that it would be an important move to describe caring science as a main focus of nursing. With an explicit origin in the world of the patients, as described above, it is obvious that it is health and well-being that must be focused upon, that caring methods must be fed by human sciences and natural sciences, and that both qualitative and quantitative methods are needed. Not least, if nursing builds upon caring science the professionalism would strengthen.

When it comes to scientific training I think that we must do much better. There are significant gaps in both "quantitative" and "qualitative" education. Beginning with the quantitative side I argue that the statistical accuracy must be greater, the database and level of analysis must be calibrated with greater precision, and a discussion of how statistical results can be interpreted for application in practice must increase. In the education of qualitative methods the high prevalence of "content analysis" must be hindered in favor of more informed and in-depth qualitative approaches. My experience is that both students and advisors tend to avoid approaches and methods that build on philosophy, e. g. phenomenology and hermeneutics. This is an instance when they throw out the baby with the bathwater. The results of this acting can be seen in the use of shallow methods that offer nothing but shallow findings. Throughout the years I have been able to see that at universities where they acknowledge theory and philosophy of science there is a more serious awareness of both benefits and weaknesses of different research approaches and methods. With more philosophical insights, the level of the critical debates are higher, the debates are more informed and the reflections more nuanced and deeper. In such milieu, the insights into what an evidence based practice means are also more well-informed even in their complexity. In such milieu, the idea of evidence grows stronger, and it gives room for the insight that it is the current research topic that must forego the design of methods, not vice versa. I also ask for reconciliation between scholars who are experts in qualitative methods and quantitative methods, respectively, so that we can develop a better combination of methods. But then, we already know that.

References

Dahlberg, K., N. Drew and M. Nyström. 2001. *Reflective Lifeworld Research*. Lund: Studentlitteratur.

Dahlberg, K., H. Dahlberg and M. Nyström. 2008. *Reflective Lifeworld Research* (2nd ed). Lund: Studentlitteratur.

Dahlberg, K., L. Todres, and K. Galvin. 2009. "Lifeworld-led Healthcare is More Than Patient-led Care: An Existential View of Well-being." *Medicine, Health Care and Philosophy* 12: 265-271.

Dahlberg, K. and K. Segesten. 2010. *Hälsa och vårdande i teori och praxis [Health and Caring in Theory and Practice]*. Stockholm: Natur & Kultur.

Ekebergh, M. 2001. *Tillägnandet av vårdvetenskaplig kunskap – reflexionens betydelse för lärandet [Acquisition of Scientific Khnowledge for Healthcare]*. Åbo akademi, Vasa: Åbo Akademis förlag

Gadamer, H-G. (1996). *The Enigma of Health: The Art of Healing in a Scientific Age*. (J. Gaiger & N Walker, Trans.). Stanford, CA: Stanford University Press.

Galvin, K., Emami, A., K. Dahlberg, M. Ekebergh, E. Rosser, J. Powell, S. Bach, B. Edlund, T. Bondas and L. Uhrefeldt. 2008. "Challenges for Future Caring Science Research: A Response to Providing Evidence for Health-care Practice." *International Journal of Nursing Studies* 43: 923-927.

Hallberg, I. R. 2006. Challenges for Future Nursing Research: Providing Evidence for Health-care Practice. *International Journal of Nursing Studies* 43 923–927.

Peters, M. A., K. Hammond and J. S. Drummond. 2007. "Gadamer's 'Enigma of Health': Can Health be Produced?" In *The Philosophy of Nurse Education* edited by John S Drummond and Paul Standish, 210-225. Basingstoke: Palgrave Macmillan.

Ranheim, A. 2011. "Expanding Caring." *Theory and Practice Intertwined in Municipal Elderly Care*. Unpublished Doctoral Dissertation. University of Linkoping.

Ranheim, A. and K. Dahlberg. 2012 "Expanded Awareness as a Way to Meet the Challenges in Care that is Economically Driven and Focused on Illness – A Nordic Perspective." *Aporia*

Todres, L., K. Galvin and K. Dahlberg. 2007. Lifeworld-led Healthcare: Revisiting a Humanising Philosophy that Integrates Emerging Trends. *Medicine, Healthcare and Philosophy* 10: 53-63.

5

John S Drummond

Senior Lecturer in Nursing

The University of Dundee, UK

---

1. How were you initially drawn to philosophical issues regarding nursing?

Following the opening remarks and style of Michel Foucault's inaugural lecture at the College de France (Foucault 1970), I would like to imagine that we had been discussing this question for quite some time, and I just happen to be the one who is currently speaking. Like others in this volume, and other volumes in the 5 questions series, I have always had an innate interest, and indeed fascination with questions philosophical. I am talking early to mid-teens when, like many adolescents, the big questions begin to kick in, or emerge into consciousness as if somehow they had always been there. Of course, this is largely abstract thought I am talking about and not philosophy per se. Still, a philosophical predisposition was there; the younger years prior to and during nursing were taken up with learning about philosophy, its history, its purpose, its diversity. My early 'schooling' was purely analytic while my latter schooling and mentorship was of a more poststructuralist leaning, an important relationship to which I return further below. I should state at this point that my approach to this chapter will be largely historical in a manner that relates to the present. I would also add that this historical approach will reach beyond nursing itself onto the wide. To save political correctness and grammatical acrobatics, I will refer to the nurse in the female. As a 'male nurse', I have no problem with this.

As regards philosophical issues concerning nursing, ironically perhaps, at first I was not particularly aware of the idea that nursing actually had any philosophical issues, nursing being a practical skill-based, and at times technology-based discipline rather than one that was research driven, or indeed theoretically embedded. But, both suddenly and gradually, over a period of twenty years, all of this began to change. This would be in the 1960s to the early 1970s through to the 1980s when novel discourses about nursing began to appear, as if from nowhere and, at first blush, for no apparent reason. They came initially from the United States of America, from emerging 'Faculty' in a search for something as yet unexpressed. It began with what was referred to then as the Nursing Process, an empirical template extracted from the accounting and costing departments of private hospitals in the US (de la Cuesta 1983). How much do we charge for assessment, how much for planning, how much for intervention, how much for evaluation? How well this compartmentalization of a costing process seemed to fit (or could be made to fit) the stages of nursing practice, and education by Faculty *for* practice, which is how it was promoted, and published in article after article (just to make sure that we all had the grasp of it). At last, here was something beyond the reproduction of tradition, this foundling, this new concept. Forth it went into syllabi as a putatively liberating shell; but liberation from what? Apparently it was a critical work of liberation from three interrelated dimensions that arguably characterized the then status quo. I will refer to them in my own words as 'the three critiques of history'. First was the critique of the sediment of history (*we have always done it this way*). Second was the critique of the 'handmaiden image' of the nurse, and third was the critique of the 'medical model' of thought.

Let me attend collectively to these three critiques without rehearsing them in detail, as there is a sense in which they contain the seeds of my answers to the four questions that follow. From a historical perspective, my first point would be that, although the three critiques are not identical, they are clearly inextricably related. It is perhaps only by looking back now, say forty to fifty years later, that we see why this is so. An appeal to the Nursing Process (so called) is clearly an appeal to the rational as a mode of thought that is independent of mere historical reproduction, an escape route then, an *entrée* into new concepts. When the nursing process hit the press, there was a not unexpected reaction from the 'traditionalists' (*we already do all this*). Probably they did, but it was held that they did it in a medical model of thought (*the unknown appendix in bed 9)*. The 'process' was no longer about medical derivations. It was about nursing, the voice of nursing seeking to speak itself to itself. Thus the putative essence of nursing was being extracted from its medically embedded history. It is at this point that I want to broaden the context, as there was, in my view, clearly something more than nursing going on. So what was it?

Let me begin with my premise that major changes in nursing structure, organization and thought tend to come from the outside, as opposed to being internally driven by the nursing profession itself. This is to say that they are also happening on the outside of nursing and enter the nursing domain, along with other domains. Take, as a pertinent example, what we now refer to as the 1960s, that decade of change and intensities.

The term 'the 1960s' is nowadays an image rather than a literal decade. It probably began in 1962, or if we go with the UK poet Philip Larkin, in 1963, "between the end of the *Chatterley* ban and the Beatles' first LP" (Larkin 2012, 90). In relation to the three critiques of history, two things emerged in the 1960s. The Nursing Process may have been an appeal to the rational, but the 1960s also saw the emergence of the post WW2 'individual'. This individual emerged from a collective passivity of the recent past; a new progressive liberalism was afoot. Thus we saw child and student-centred education, and patient-centred nursing, not to mention person-centred counselling as a dimension of Holocaust psychology. The 1960s was also a time of what is referred to as 'second wave feminism' where the nurse emerges as more than the handmaiden to the medical man. This was a glorious age when anything could come or go, until the 1960s came to an abrupt end in 1973/74 with the Arab oil embargo, due largely to the side of the West with Israel in the *Yom Kippur* war.

However, in nursing, a legacy remained and was developed. The Nursing Process could not simply be an empirical shell, a hollow abstraction. Something had to be assessed, planned, justified, implemented, evaluated, whether by linear means or by recurring and omnipresent spirals, often of a spectacular nature. Through the conceptual idealism of the nurse-patient relation and patient-centred nursing, there emerged the 'individual care-plan'. The journals of that time abounded with examples of care-plans. I cannot cite them all here, but they were not too far removed from the exemplars given at: http://www. rncentral. com/nursing-library/careplans. Point is, the care plan had to be written down in its totality, like a scripture. Care plans lay in drawers or filing cabinets like the literary side of a parallel universe. There was the act of nursing and the writing of nursing. It was as if nursing was writing itself out of the past, if not without a liberal helping of scepticism. The mid-to-late 1970s were dark and confusing days. Everyone was in on the act. The 'individual' was being invested in.

This investment, it should be said, did not only include the patient. It also included the nurse. After all, if the nurse was to be liberated from the medical model of thought, answers needed to be proffered as to the difference between the two professions. In other words, what *is* nursing exactly if not an adjunct to medicine in general terms? What is *unique* to nursing? Or more elaborately, what is it *about* nursing that makes it *nursing* as opposed to something else? In what was a time of change and reflection, this was a not unreasonable question to ask. Looked at philosophically, or at least analytically, it does (and did) however contain elements that render (and rendered) it a difficult question to answer in simple terms. The first element is that the concept is *inhe-*

rently polyvalent. It is used in everyday language, not just professionally. Add to this that these everyday uses may differ quite dramatically (*the nursing mother; she was nursing her wrath; it was skilful nursing that pulled her through*). I don't want to wander into Wittgensteinian language games; suffice to say that we are looking here at family resemblances rather than the lure of an essence that would cover all instances of the way in which the word is used, even, I would suggest, professionally. The second element is that the concept of nursing is also *strategically* polyvalent. By this I mean that, even within the profession of nursing itself, people can invest in the concept for entirely different reasons. At the time in question (we are now in the early 1980s) such is exactly what happened. The individual care-plan, designed around the sequential stages of the nursing process, needed, or was held to need an overarching model that captured the nature of nursing itself. Thus came Nursing Models. Why they were called 'models', I have no idea, these Panglossian calls to both perspective and essence. This was followed by something called 'Nursing Theory', which overlapped with Nursing Models. Finally, another development in the 1980s was that of the 'primary nurse'. This was the apogee, the climactic professionalising moment. The primary nurse would, in theory, be allocated a number of patients and would carry 'primary' responsibility for the process of assessment, planning and so forth according to the care plan. In a nutshell, the primary nurse with her process and her care plan designed around a particular nursing model would be exercising a new form of professional autonomy (Primary Nursing, 2012). Yet there was something wrong.

It was all of this that drew me into philosophical issues regarding nursing. I will not take the high ground. I will not rant. I will instead return to the poet Philip Larkin (2012, 60). In his response to the question "What are days for?" we might well substitute the question 'What is nursing for?'

> *Ah, solving that question*
> *Brings the priest and the doctor*
> *In their long coats*
> *Running over the fields.*

And so it was. Suffice to say that something appeared to have gone seriously wrong, had gone awry between models that spoke of goals of attainment, or adaptation, and others that spoke of units of energy on the space-time continuum, and others yet again of activities of daily living that failed to capture an essence that was always polyvalent. It is easy to look back now, but of course there was no essence. I thought of parsimony. I thought of Ockham's razor. I scribbled some notes but did not seek to publish anything. I was still too busy thinking and trying to come to terms with my apparent ignorance and incredulity both together. This takes me to the next question.

2. What, in your view, are the most interesting, important, or pressing problems in contemporary philosophy of nursing?

In responding to this question, I want to continue with the historical narrative. The changes I have outlined above were at the time often referred to collectively (if somewhat loosely) as 'the new nursing'. It is not without import that the immediate post-Nightingale nursing in the late 19th and early 20th centuries was also referred to as the 'new nursing', with the pre-Nightingale nursing being referred to as an old-style nursing. So here in the 1980s we had another new nursing with the sublation of 'the three critiques of history' into a Nightingalesque old-style nursing that must be cast off like a skin that no longer absorbs the light. This new nursing of the 1980s had its virtues. It was in some respects a major step forward at a certain point in history. Something had been loosened from the grip (the manceps). Yet there will always be a tendency to the manceps. What is the true manceps? It is highly significant that the new nursing did not become established as an internally driven force; perhaps it emerged as such in discourse, but it did not do so in practice. This was because something else that was new also emerged, and again from the outside of nursing. It was what is commonly referred to as the 'new managerialism'. The new managerialism became the new manceps – the new grip, a new form of assimilation. This is to say that it transformed the new nursing's professionalising thrust by a subtle yet colonising shift from what was purported to be an autonomous professionalization into a different form of the technorational. The care plan became part of the audit trail. The primary nurse became 'the named nurse' and thus an object of accountability. What we were seeing here was a new discursive form of bureaucratisation, of efficiency, yet also of surveillance. Of course, it was not all a one-way street, but it was tensive, and worthy of note. The new managerialism shifted discourse in ways that still exist today. Problems became 'challenges'; loss became an 'opportunity' for something 'new'; everything became 'exciting'. A business model was slowly inserting itself into healthcare consciousness – providers and users. A new tradition had (and has) established itself in the breeding grounds of the Academy and the Clinic, in their auditoria, in their warrens and their hives.

To broaden this onto the wide, the mission of the new managerialism (based partly on Thatcherism (UK) and Reaganomics (US)) was to 'roll back the state' in the interests of private enterprise. This involved a climactic battle with the trade unions on many fronts as heavy industry went into decline – steel, coal, cars, and ships. World politics and commerce was also changing as manufacturing moved slowly eastwards where the wages were cheaper and the unions less strong if they existed at all. What we are talking about here is the emergence of a new market economy, its hunger for the dollar, and its liberation due to advancing computer technology. It is important to see the changes that were going on. I would see the notion of 'the changes that are going on' to be the most pressing or important problems in contemporary philosophy of nursing. I would express it, after Deleuze, as 'a sensitivity to signs' (Deleuze 2008). Take for example, the sudden shift of the care of the elderly into the independent sector, into spaces at the edge of the city (the *citivas*), these colonies of frailty. Money moves in and money moves out. Nurses work their shifts as money is shifted across continents in the blink of an eye. It is this that I would pay attention to. I could go on (Peters and Drummond 2008).

3. What, if any, practical and/or socio-political obligations follow from studying nursing from a philosophical perspective?

Beyond the obvious obligations of open-mindedness and honesty, I would add that it is also about asking the right questions that extend beyond nursing itself (my theme of always looking to the outside). Why are many drugs so expensive? Why is medical equipment so expensive? Why, when a new form of treatment is pronounced, is it also pronounced that it may take at least another ten years before it comes on to the market? What market? Why are countries and healthcare opportunities brought to their knees by venture capitalism (Drummond 2001)? These are complex questions, but it is important to keep asking them, to keep chipping away. I would also like to add that philosophy of nursing is not just about the patient, the client, the receiver of healthcare. It is also about the nurse. There is an obligation to the practising nurse, an obligation that must be devoid of patronage, and of stipulation (Drummond 2000).

4. In what ways does your work seek to contribute to philosophy of nursing?

I am at heart a poststructuralist thinker, and it is this concept of poststructuralism that I wish to expand upon. For a start, the term poststructuralism is often conflated with postmodernism, or the postmodern. This, in my view, is a profound error; an understandable and not uncommon error perhaps, but still clearly erroneous, not to mention unhelpful as used by many nursing (and other) authors. Let me explain.

Postmodernism is a socio-cultural phenomenon, not a philosophical approach. Poststructuralism as a philosophical approach does not reject the analytic or indeed the Post 'Enlightenment' tradition of the 'West'. Rather it interacts with it, not to seek its demise but to move the work of thought forward in an ongoing and reciprocal dialectic. In broad brush terms, this relates to what is often referred to as an engagement with 'the project of modernity'. This dialectic is not Hegelian in the populist misinterpretation of the Spirit of History. Yes, it is reciprocal, but it is also asymmetrical. By this I mean that the relation between poststructuralism and the analytic tradition is not one of mutual exclusion. Neither is poststructuralism simply a diverse bundle of negative critiques. I tried to demonstrate this in my paper 'Nursing and the avant garde' (Drummond 2004) as well as the two citations in the previous question above. There have also been other publications in which I try to demonstrate that poststructuralist approaches to philosophical enquiry can be an experimental and progressive undertaking and not some postmodern fantasy (Drummond and Themesslhuber 2007). Poststructuralism is not a singular or even broadly unified church (see Williams 2005). Neither is it a rampant advocate of relativism (Drummond 2005). It is in this respect that I have tried to introduce the works of major poststructuralist philosophers in a positive way, and in a manner that blurs the often overstated distinction between the analytic and the continental traditions in philosophy. If my approach differs from others, I tend to write on behalf of the nurse.

5. Where do you see the field of philosophy of nursing to be headed, including the prospects for progress regarding the issues you take to be most important?

I do think that philosophy of nursing is emerging as a bona fide discipline, but the emphasis is on the word 'emerging', and in ways that may be difficult to predict. For example, there is the question of the relation between philosophy of nursing and philosophy of medicine. To this we could add the relation with healthcare ethics, including nursing ethics. All of these operate on the same broad landscape where the difference of emphasis is not always clearly identifiable. Perhaps philosophy of nursing is more catholic in its embrace of diversity. Although every philosophical project arguably contains an inherent moral dimension, the range of papers in various journals, and in particular *Nursing Philosophy* and *Nursing Inquiry*, is very promising. There is also *Aporia: The Nursing Journal*, which is quite unique in that it has papers published in English, and different papers published in French. We have moved beyond a dominance of ethics *qua* ethics to a range that includes 'the body', the nature of spirituality, women's studies, political critiques, the nature of palliative care; the elderly, the child, the state of practice organisation (the clinic), and the state of the academy to name but a few. We could say that philosophy of nursing is a comparatively recent development compared say to nursing ethics which is well established, and also with its own journal of the same name in the UK. That said, the discipline is gathering momentum and philosophical papers are now accepted by most of the leading nursing journals with an international profile and readership. As noted above, there is now an International Philosophy of Nursing Society (IPONS), founded in 2003. IPONS has its own journal *Nursing Philosophy*, published by Wiley-Blackwell, along with an annual international conference, so philosophy of nursing is out there. I think it important that nursing as a worldwide profession embraces this philosophical dimension. We should of course acknowledge that philosophy, including applied philosophy, is not everyone's 'cup of tea', but it is my view that it should be available to all, even if they just happen to come across it.

As I come to a close, there are various things I want to follow through. The first is that philosophy of nursing is not to be conflated or identified with what I referred to above as 'nursing theory'. Second point is the question, what is philosophy of nursing out there for? A more direct way of asking this question might be 'what does philosophy of nursing contribute to practice?' If this is a difficult question to answer it is because, at first blush, the answer appears not to be obvious. But this may be because we treat the question superficially, in which case two basic slippages may occur. The first would be to conceive of philosophy (including philosophy of nursing) as 'theory' when, in fact, it is about different forms of analysis applied to different types of problems that *concern* practice. Connected to this is the second slippage in which we may conceive of practice as only clinical practice, and clinical practice only. In exploring the question it would be more productive to conceive of 'practice', or even the possibility of practice, as multidimensional – clinical, education, research, administration, governance, politics – all is practice. Even thinking is a practice. It would be then that ideas and apologies begin to emerge.

That said, I do think that nursing needs a degree of analytical protection from that which roams away from what nurses actually do. At times it is useful to remind ourselves that philosophy, although it serves many purposes, was originally (and hopefully remains) about what constitutes a good life. I would also like to see nursing philosophy connect more with science in all its relevant dimensions.

And so it is, here at the end, having been the one who is currently speaking, there is a sense in which I wish this discussion was just getting off the ground. There is much yet to be said. After all, at the day's demise, when everyone has gone home, and the last person to leave the building turns off the lights and locks the doors, there is always one who remains – that iconic figure, that timeless image - the nurse. She will walk the wards; she will attend to needs; she will sit with her writing at the desk or read the monitors for the latest reports. In the dead of night she will cast away her notes when the arrest bell rings. She is the object of politics (often crass and ill-informed), of theory, of bad press and good press. Yet still she remains. Philosophy of nursing, it is for her, for nursing itself. We must continue to make it so as nursing continues to change. There will always be a new nursing. There will always be a new manceps. Bring it on.

References

de la Cuesta, C. 1983. "The Nursing Process: From Development to Implementation." *Journal of Advanced Nursing* 8(5): 365–371.

Deleuze, G. 2008. *Proust and Signs*. London: Continuum Impact.

Drummond, J. S. 2000. "Nietzsche for Nurses: Caring for the Übermensch." *Nursing Philosophy* 1(2): 147-157.

Drummond, J. S. 2001. "*Petit Différends*: A Reflection on Aspects of Lyotard's Philosophy for Quality of Care." *Nursing Philosophy* 2(3): 224-233.

Drummond, J. S. 2004. "Nursing and the Avant-garde." *International Journal of Nursing Studies* 41: 525-533.

Drummond, J. S. 2005. "Relativism." *Nursing Philosophy* 6(4): 267-273.

Drummond, J. S. and M. Themesslhuber. 2007. "The Cyclical Process of Action Research: The Contribution of Gilles Deleuze." *Action Research* 5(4): 430-448.

Drummond, J. S. 2007. "Care of the Self in a Knowledge Economy: Higher Education, Vocation and the Ethics of Michel Foucault." In *The Philosophy of Nurse Education* edited by J. S Drummond and P. Standish, 194-209. Basingstoke: Palgrave Macmillan.

Foucault, M. 1970. *The Discourse on Language*. Inaugural lecture delivered at the College de France. Available at: http://www. wordpress. com/2012/05/08/the-discourse-on-language Accessed14th September, 2012.

Foucault, M. 2001. Madness and Civilisation. London: Routledge.

Larkin, P. 2012. "Days." In Philip Larkin: The Complete Poems edited by Archie Burnett. London: Faber and Faber. Also available at:

http://www. poetryfoundation. org/poem/178046
Accessed September 10th, 2012.

Larkin, P. 2012. "Annus Mirabilis." In Philip Larkin: The Complete Poems edited by Archie Burnett. London: Faber and Faber. Also available at: http://www. wussu. com/poems/plam. htm Accessed September 14th, 2012.

Peters. M. A. and J. S. Drummond. 2008. "Political Economies of Health: A Consideration for International Nursing Studies." Policy Futures in Education 6(3): 351-362.

Primary Nursing. 2012. See:
http://www. google. co. uk/
search? q=Primary+Nursing&rls=com. microsoft: en-gb: IE-SearchBox&ie=UTF-8&oe=UTF-8&sourceid=ie7&rlz=1I7RNRN_en&redir_esc=&ei=TGNTUKOsJMP80QXYs4CQDw

Williams, J. 2005. Understanding Poststructuralism. London and New York: Acumen Publishing.

6

Katie Eriksson

Professor in Caring Science

Department of Caring Science

Åbo Akademi University, Vaasa, Finland

---

1. How were you initially drawn to philosophical issues regarding nursing?

My interest in philosophy, history – especially history of ideas, and ethics was present early on in my life. In upper secondary school (*Swe: gymnasium*), I chose to study philosophy, and a visionary teacher awakened my interest in the classical Greek thinkers Plato, Socrates and Aristotle. During my studies to become a nurse (1962-1965), I had the opportunity to delve into nursing history and especially the ethical questions alongside the regular courses. This experience allowed me to form a new way of thinking in relation to my work. In Florence Nightingale's thinking, the history, practice, ethics and belief in the future are merged. This, together with the thoughts of baroness Sophie Mannerheim – the Finnish pioneer within nursing education, allowed for a different view of the nurse to evolve, where the importance of high levels of ethics, proper leadership and theoretical knowledge, is emphasized, along with a strong devotion toward the patient. As a young, newly graduated nurse, I came to work at an emergency ward in my hometown. I often had the night shifts and as there was periodically time left over for other pursuits in addition to the nightly routines, I decided to enrol in a correspondence course (*Swe: Hermodskurs*) in the history of philosophy. I was able to deepen my knowledge of Plato, Socrates and Aristotle's thinking. These philosophers have since stayed with me throughout my life in various ways.

During my studies at the University of Helsinki, in the Faculty of Philosophy, I met Professor Peep Koort, who at the time was a newly appointed Professor of Pedagogy and he became my mentor and friend. Peep Koort was a visionary and he wanted to create depth and new thinking, ideas and new approaches within the field of education. Peep Koort was ahead of his time, and during the 1970s he had already developed a model of concept definition, semantic and configuration analysis based on hermeneutics. It was through him that I came into contact with conceptual thinking and the first work I did along these lines was my licentiate thesis (1974), which delved into the concept of health from a conceptual philosophical perspective. An important finding from this study was the 'health cross', describing the multidimensional aspects of health, which resulted in the book *The Idea of Health* (1987). The concepts here were an entirely new way of thinking within the field. Within me, a vision of a Caring Science as an academic discipline concerned with human being and caring with a more fundamental meaning, in addition to the significance of various concepts which could open up windows toward a new world, and a new reality could begin to unfold. The work with concept development and with the field's core concepts (suffering, health etc.) still sets our department apart from others in the field.

It is often special encounters with people that are of importance when it comes to how things turn out. During a visit to the University of Gothenburg, I met the newly appointed first Professor of the Theory of Science within the Nordic countries, Håkan Törnebohm. During the 1980s, he was an integral part of laying the foundation for the theory of science for caring science. He introduced the concept of paradigm and suggested that all sciences and professions have their own paradigm. Many questions regarding philosophy of science were discussed with him, concerning paradigm, ontology, views on human being and on views on science.

In the middle of the 1980s, the writing of teaching material and aids concerning substantive issues within caring science became of immediate interest. Plato's teachings about ideas and the thoughts of Aristotle became a natural starting point. *The Idea of Caring* (1987) became an important fundamental book within the field, where the substance of caring was described on the supposition of a foundation consisting of the history of ideas, conceptual and philosophical aspects. The substance of caring was identified as tending, playing and learning, along with faith, hope and love. As there were no developed models for a systematic science of caring, a research program based on our view of caring was needed. In the book *The Pause* (1987), the guiding principles were laid out for the first research program based on the thinking about the idea of caring with the focus on basic research.

In the forming of Caring Science as an academic discipline, the theologian Anders Nygren's writings, especially "Eros och Agape" (Author's translation: Eros and Agape) and "Mening och meteod" (Author's translation: Meaning and Method) became pivotal in serving as guidance for the thoughts on theory of science and science of philosophy. Nygren's idea on basic-motive (*Swe: grundmotiv, Du: grondmotif*) and on contexts of meaning as having an inherent order in terms of scientific knowledge were of great importance for our work.

The ontological question, that is the questions of what the inmost being of caring reality actually is like, came to the fore in our search for knowledge. It is the question of 'what' rather than 'how' that leads to the scientific formation of concepts and theories. In our search for the fundamental category of caring, that is, the basic concept or the idea behind all forms of caring, we have used different approaches such as historical research and motive research (Nygren 1972). From the standpoint of concepts and the history of ideas, two leading conceptions of caring come to the fore: that of compassion and that of human love, i. e. the caritas motive, which we see as the basic motive of caring. The caritas motive at the same time expresses the basic value of caring science and the ethical inducement to all forms of caring. The leading idea of caring is to alleviate human suffering and to preserve and safeguard life and health.

The meeting and encounters with Kari Martinsen over thirty years has, through our dialogues about the 'big questions', had great significance, above all for the development of substance. Through these dialogues, Anders Nygren's and Knud Løgstrup's voices have similarly been heard. During the past years, Hans-Georg Gadamer's philosophical hermeneutics has constituted an important foundation for the continued development of a caring science based on hermeneutics and with a praxis that results in a way of working with an inherent dimension of art. For example, even though you meet the same patient for the tenth time, there is something new, professional, yet personal in every meeting.

2. What, in your view, are the most interesting, important, or pressing problems in contemporary philosophy of nursing?

One of the most important philosophical questions today regards the continued development and preservation of an autonomous Caring Science based on the human sciences. Another important question is the view of Caring Science as a scientific discipline and an independent subject as a basis for nursing and the nursing professions. This additionally concerns the continued development of the theory of science for caring science, which I see as a wider application than philosophy of nursing. Theory of science involves ethos and ethics, ontology, epistemology, methodology and the development of a concept of evidence of human science.

Another pressing matter concerns protecting a tradition, which has been developed for millennia, to be handed over, and make evident, i.

e. to make visible and verbalized in time. A merger between the past, the present, and visions of a future is a bountiful foundation within caring. It is by rediscovering the fundamental core concepts, such as love, mercy, dignity, suffering, we are able to arrive at a deeper insight into the reality, which conceals itself beyond the physical and what is immediately evident. Scientific language is slowly but surely being depleted and is increasingly becoming lifeless and technical. Through concepts and words that are carriers of meaning and significance, and ethics that describe human being in all of its shifting possibilities and conditions, you can provide for a different breeding ground for caring and nursing.

A current issue of importance involves the development of methodologies for a human science research within caring science. It is primarily methodologies in a hermeneutic spirit that are needed as an alternative to the flow of various methods that reduce a great deal of the substance of caring. Such methodologies are hermeneutical reading, textual interpretation and hermeneutical observation along with other empirical methods. The basic research within science of caring composes the basis of all meaningful clinical research and should have a greater deal of attention among researchers.

A thorough philosophical discussion of the theory of science within caring science from a human science perspective should be held. A consequence of the lack of this type of discussion could potentially mean a reduction of the meaningful caring substance, if we continue to submit ourselves to a narrowly defined scientific concept of evidence. Kari Martinsen and I have in our book *To See and To Realize* (*No: Å se og å innse*) tried to open this issue up for discussion. The ever-present and ever-lasting core concepts are what create an inner discipline to knowing and constitute an ontological boundary as a prerequisite for contextual evidence.

3. What, if any, practical and/or socio-political obligations follow from studying nursing from a philosophical perspective?

The most important issue is the mission and position of the universities and of science in an increasingly de-humanized society. Caring science as a human science must be provided a space within the universities and the universities of applied sciences. As a higher education organization has a responsibility as carriers/upholders of culture, caring science must be given a voice and to become more visible in compliance with the Humboldtian notion of education, which constitutes the corner stone of the Nordic universities and universities of applied sciences.

Through the development of scientific thinking, new possibilities and innovations are made possible. To me, theory and practice form a unity. The boundaries of reality are widened and deepened through a constant motion between *vita activa* and *vita contemplativa*, the active life and thinking. Since 1996, I have had a minor appointment, alongside my professorship in the Department of Caring Science at Åbo Akademi University, as a head nurse at the hospital district of Helsinki and Uusimaa, Finland. This has given me the opportunity to see how research, science and practice can be interwoven to form a whole. In order to achieve this merge, some prerequisites include recognizing the "common", to find a common foundation of values, to recognize and to implement an ethos that unites and that all can embrace whole-heartedly.

In my work, together with professional carers in various development projects, I am inclined to posit that people in the field are in no way unfamiliar with theoretical and philosophical questions. My model of the caring process was integrated in the work of several clinics in the beginning of the 1980s. Today, the development of care has been concentrated around the idea of caring, suffering and health to mention but a few examples. I believe that there are many presuppositions and structures in organizations, which hinder the professional carers from developing independent thinking and from practising the art of caring in patient care.

4. In what ways does your work seek to contribute to philosophy of nursing?

Throughout my work that started in the beginning of the 1970s when I worked as a teacher and principle at the Helsinki Swedish College of Nursing Education, my work in itself has been an on-going contribution to the development of a philosophy, a caring thinking as a basis for nursing. In the middle of the 1970s, the issue of caring science and the nurses' right to carry out research was raised. This constituted the first contours of the caring science on a human science basis that today forms the tradition at Åbo Akademi University.

In the middle of the 1980s, chairs where established in five of the Finnish universities, and I received the unique possibility to develop and, entirely from the ground up, shape this new field of science, Caring Science, and to establish a Master's and PhD research education. Research and education in caring science at the Department of Caring Science at Åbo Akademi University in Vasa, Finland started on September 1st, 1987. I have had the great pleasure and honour as a Professor of Caring Science to work for 25 years together with Unni Å. Lindström, Professor of Clinical Caring Science, and a knowledgeable staff in developing caring science. Unni Å. Lindström has done pioneering work in the development of clinical caring science as an academic discipline and an evidence based caring praxis. Our department has a Nordic profile with postdoctoral candidates from all the Nordic countries, and an extensive post-doctoral network. It is through all of this that knowledge and value transfer can take place and all of our doctors and masters degree students become messengers of the caring tradition around the Nordic region on various posts within hospitals and the health care sector.

Our work with the development of a caring science, caring as a core concept and the caritative caring ethics based on an ethos, I see as the most important contribution to a philosophy of nursing. The caritative caring theory, which I developed during the 80s and early 90s, has become a basis for nursing internationally, further contributing to the field. Through this work, and the work of our department, I am able to discern that we have been able to contribute to the development of a philosophy, a theory of science for caring and nursing in a very tangible fashion. Through basic research we have been able to develop the clinical research and to give it a distinct profile. It is through graduated Master degree students and Doctors, and by publications, dissertations and the participation in discussions on various levels that we can continue to contribute to the philosophical discussion.

5. Where do you see the field of philosophy of nursing to be headed, including the prospects for progress regarding the issues you take to be most important?

There are few trends within today's caring science. On the one hand, there is a move toward an increased fragmentation due to increased multiplicity. On the other hand, an increased interest in the great fundamental philosophical questions within caring can be discerned. I believe that the multiplicity of knowledge and technological challenges are constantly increasing in numbers and instances, but the problems inherent in this development can be counteracted by constantly holding the fundamental philosophical questions alive by allowing space for contemplation, reading and lingering before the reality we can sense beyond the immediate.

Basic research within caring science is on the brink of the task of organizing the multiplicity of various outer structures, constructions of methods that are a consequence of the modern science anchored in an outer logos, which has concealed the fundamental fragments and the fundamental order within Maya's veil. It is important to rediscover the unity in the multiplicity. According to Cusanus, unity stands for truth, the good and the beautiful.

The great philosophers of the past still have much to tell us, and their thoughts, ideas and questions are current even in our time. Philosophy and the history of ideas need to receive a greater emphasis and attention within all levels of education. Philosophy, love for wisdom, science, truth and the good, will always have a place in every human's life.

Selected Works

Eriksson K. 2002. "Caring Science in a New Key." *Nursing Science Quarterly* 15: 61–65.

Eriksson K. 2006. *The Suffering Human Being*. Chicago: Nordic Studies Press. [English translation of: *Den lidande människan*. Stockholm, Sweden: Liber Förlag. 1994.]

Eriksson K. 2010. "Evidence: To See or Not To See." *Nursing Science Quarterly* 23: 275-279.

Eriksson K. 2010. "Concept Determination as Part of the Development of Knowledge in Caring Science." *Scandinavian Journal of Caring Sciences* 24: 2-11.

Eriksson K and UÅ Lindström. 2007. "Vårdvetenskapens Vetenskapsteori på Hermeneutisk Grund – Några Grunddrag." [Hermeneutic Basis of Health Sciences' Scientific Theory - Some Basic Features] In *Vårdvetenskap och Hermeneutik* [*Caring Science and Hermenutics*] edited by K Eriksson, UÅ Lindström, D. Matilainen and L Lindholm, 5-20. Vaasa: Division of Health Sciences, Åbo Akademi University.

Lindström UÅ, L. Lindholm L and JE Zetterlund. 2010. "Katie Eriksson: Theory of Caritative Caring." In *Nursing Theorists and Their Work* 7th ed. edited by A Marriner Tomey and M Raile Alligood, 190-221. St. Louis: Mosby Elsevier.

Martinsen K and K Eriksson. 2009. *Å se og å innse*. [*To See and To Realize*]. Oslo: Akribe.

7

Sally Gadow

Professor Emerita

College of Nursing, University of Colorado, Denver CO, USA

---

1. How were you initially drawn to philosophical issues regarding nursing?

During my masters program in nursing, I realized that nursing needed a philosophical foundation. I wanted more fundamental answers to "What is nursing?" than the menu of available answers: healer, helper, health educator, consumer advocate, counselor, parent surrogate, physician associate, contracted clinician. A sociological description of roles performed by nurses failed to identify an essence, so I detoured into a doctoral program in philosophy, then returned to nursing with the hope of elucidating a philosophical core.

2. What, in your view, are the most interesting, important, or pressing problems in contemporary philosophy of nursing?

For me the most pressing problem has been how to understand and describe the self existentially as embodied, constructed, and relational.

Understanding the self (nurse as well as patient) as essentially embodied is a postdualist approach, leaving behind the rationalist dichotomy of mind versus body and embracing instead an existential view. My approach has been dialectical, elaborating a series of self-body configurations that affect health. Three of these are the lived body, object body, and subject body. Without distinguishing at least these three in a clinical situation, an ethical problem such as how to support patient autonomy is reduced to a mechanical procedure that has little bearing on the patient as a complex person.

For example, the ethical question of whether to support a patient's refusal of treatment because of the pain it causes cannot be answered without understanding embodiment. Insisting on the body as only object denies its lived reality; conversely, allowing the subjective immediacy of pain to dictate a refusal denies the value of objective knowledge. Both denials create a chasm between a patient overwhelmed by the body and nurses who, in order not to suffer with the person in pain, must dissociate entirely from their own bodies. Ethical inquiry that omits the ambiguities of embodiment can address only a shadow of human experience.

Analyzing embodied experience leads to the second problem that I see as fundamental: the constructed self. While embodiment across individuals may share the phenomenological forms I identify, each person creates an individual version of those. That version can be considered a personal narrative. Because it is not composed in a vacuum, general cultural narratives influence its development. Hermeneutic analysis can identify the social, clinical, political, and ethical narratives out of which personal narratives are constructed. Without an understanding of the dominant (often contradictory) general narratives that influence nurse and patient, autonomy is viewed too simplistically. Respect for persons is reduced again to a mechanical procedure that assumes *a priori* agency instead of acknowledging ambiguity.

The third problem I see as central is the relational nature of the self. Rationalist views of the individual consider a person self-contained and independent. Phenomenologically, however, individuals are intractably connected. Given the vulnerability of those with compromised health and the power of those who care for them, it is ethically crucial to understand the self as relational. We know too well the possibility for paternalism created by the meeting of power and vulnerability. It is important to identify forms of power within the experience of illness and forms of vulnerability within the phenomenon of caring. One ethical implication of the relational view is that neither nurse nor patient alone may have sufficient moral autonomy to unilaterally construct an ethical narrative about the good in their situation. Only together will they accomplish that. Their creation becomes a relational narrative, and moral autonomy becomes a function of engagement rather than independence.

3. What, if any, practical and/or socio-political obligations follow from studying nursing from a philosophical perspective?

An existential view of the self as embodied, constructed, and relational entails an ethical obligation for nursing at least to accommodate, if not actively promote, engagement between nurse and patient as the form that respect for persons takes. The engagement that I describe as relational narrative cannot be mandated; it can only be voluntary, freely entered by both sides. It cannot even be modeled, since it must be constructed differently in each situation. But it can be encouraged instead of marginalized. Policies regarding consent can require at least that the elements of a relational narrative be in place for consent to be informed. Those elements include subjective as well as objective understanding by both nurse and patient, as well as awareness of the cultural narratives that influence them.

A corresponding obligation requires nursing education to prepare students for an existential practice, giving them sufficient liberal arts foundation that they can consider engagement an option in their practice and imagine the unique forms it might take. Imaginative literature seems more vital to that preparation than formal philosophy, which traditionally creates impersonal distance. Literature portrays persons in their rich concreteness, compelling engagement with fictional characters by evoking empathy for their condition. Literary encounters with an experience such as suffering can better equip students for engagement than the study of metaphysical systems.

4. In what ways does your work seek to contribute to philosophy of nursing?

My contributions have attempted to illuminate several areas of nursing and health, including chronic illness, aging, pain, women's health, prison health care, clinical assessment, the natural environment, and interpretive inquiry. In this section I summarize my work in two philosophical domains that intersect each of those areas: dialectic and ethics.

1) I use the method of dialectic not only to distinguish opposing elements of a phenomenon or concept and to propose a synthesis, but also to demonstrate that the elements entail as well as oppose each other. Including both sides is vital to the synthesis, which in turn generates its own new opposite. Each successive synthesis carries within it as reconciled all earlier oppositions. Nursing would be diminished by the absence of any of them.

Levels of opposition and synthesis are more like regions of experience inhabited by nurses and patients at different times than stages leading to a final, supreme level. However, it is convenient to describe them as levels in order to make clear the logic of their connection, the way in which one can lead to another in a conceptual progression.

In taking a dialectical approach to a complex phenomenon, I identify as the first level the aspect of greatest immediacy, then describe each further level by looking for a construct that balances and complements the limitations of the prior one. Starting with simple immediacy, levels of increasing complexity can be identified by extending the analysis as far as imagination allows. For some philosophers, a dialectical progression finally culminates in a natural end or *telos*. For me, the pull of a *telos* feels coercive. Without assuming an inevitable conclusion, dialectic can proceed indefinitely as a method for introducing ever more possibilities. To illustrate, I describe briefly my approach to embodiment.

The first level is the sheer immediacy of the lived body, in which there is no self-body distinction; body and self are experienced as one. Few of us can recover that immediacy, because it leads directly to the next level, the object body. When the lived body is disrupted by constraint that it associates with itself rather than the world around it, body and self divide; the body becomes the other. That feeling of being encumbered gives rise to an opposition in which self and body each struggle for mastery in an attempt to restore unity.

One form of mastery by the self is reduction of the concrete body to abstract concepts. The self regains control by relegating embodiment to clinical categories. The body as conceptual object offers the self a means of reconciling with it as instrument instead of antagonist. That instrumental relationship I call cultivated immediacy, because the new unity depends on the body's compliance being so complete that it drops out of awareness, but only after great effort on the part of the self. Rehabilitation professionals work to cultivate that unity in which the self achieves its aims through an objectified body that has recovered a measure of immediacy.

The distinction between self and body so far has been described as opposition, and reconciliation has occurred in the direction of the self mastering the body. A different reconciliation can be imagined, in which the subjectivity reserved for the self is recognized in the body as well. This level I call subject body. From the perspective of the self mastering the body, illness and aging have the character of insurgence: the body rebels, the self is thwarted. If we suspend that view, a different unity can be imagined, in which the body insists, not that the aims of the self be surrendered, but that its own reality and values be accommodated. Recognizing that insistence as valid means recognition of the body as subject rather than object. The self-body relation becomes one of intersubjectivity, a partnership rather than mastery.

The experience of embodiment includes all of these levels and no doubt more. The dialectic posits no privileged level or sequence but rather suggests that experience would be impoverished without all of them. Here the language of regions instead of levels is helpful. Some regions of embodiment are more inhabitable than others in certain circumstances. Moving from one region to another as needed requires only a realization that others are available. Nurses are uniquely positioned to foster that realization, and dialectic provides them with a method for locating more hospitable regions.

2) Another way in which I have offered a contribution to nursing philosophy is through my work in ethics. A complete philosophy of nursing requires an ethical framework. I have developed a dialectical framework that includes opposing ethical approaches: immersion, detachment, and engagement, which I designate respectively as premodern, modern, and postmodern. As in my discussion of embodiment, it

is useful to analyze these as levels, then turn to the language of regions.

The ethical level I designate as premodern is the immersion in a community from which certainty arises. Unquestioned allegiance to the values of a group — family, religion, culture, profession — provides a certainty that needs no defense; a nurse knows the good without recourse to reflection. The strength of immersion as an ethical region in which to practice is the solidarity it offers. Nurses and patients sharing an unquestioned view of the good are on common ground.

That immediacy would be interrupted by the detachment required to explain or defend. Moral doubt undermines immediacy. If doubt is expressed, two responses are possible. One is to reinforce certainty by using force (shame, exclusion) to restore unity. The other is to address the question instead of silencing it, but that alters immersion irrevocably. It requires the questioner to move out of subjective certainty into an objective realm where reasoning occurs.

A single question can accomplish the dialectical turn to detachment that is the opposite of immersion. The detachment I term modern ethics is the attempt to find certainty in rational principles, codes, and laws. Because subjective immediacy is an extreme approach, its opposite too is extreme: an objectivity untainted by personal preference or group allegiance, a commitment to universal reason.

Universal principles of equality and fairness may seem an advance over a group's belief that only its own views are valid. But reason too discriminates; it favors those who transcend the non-rational parts of existence. The hegemony of reason privileges those whose situation (wealth, gender, health) affords them the luxury of transcendence as long as others remain bound by conditions of hunger, exhaustion, or pain. The metaphysical dualism in which rationality dominates is reproduced as an ethical dualism in which an experience such as suffering that defies rational control is discounted.

The dualism within modern ethics can make it an inhospitable region, since emotion, imagination, embodiment, and empathy are considered inferior to rationality as sources of certainty. Worse than inferior, they interfere; objectivity is the ability to detach from them in order to follow abstract principles without reference to human differences. Another region is needed in which persons can be valued in their diversity. That region I characterize as postmodern, where abstract reasoning is but one avenue of ethical understanding, alongside its opposite, engagement.

The move from detachment to engagement is a turn toward contingency instead of certainty, contextual instead of fixed meanings, and relentless reinterpretation because no version can be the last; other views can always be imagined. Unlike the premodern and modern regions, the postmodern foregoes appeal to an ultimate authority; an ethical narrative about the good is viewed as a human construct and subject to continual reconstruction. Personal narratives replace community allegiance and universal principles.

In illness and vulnerability an individual's capacity for maintaining a narrative may be disabled. In extreme cases, personal narratives can fail completely, leaving patients ethically adrift with no way to envision a good for themselves. At those times a new narrative is needed, and help may be needed to compose it. Engagement is the avenue for that help. Ethical narratives created by patient and nurse together are more than individual accounts; they are relational narratives. (I describe that engagement as existential advocacy in my early work.)

A relational narrative expresses intersubjectivity, an alternative to both the subjectivity of immersion and the objectivity of detachment. Without either of those sources of certainty, the narrative is located in contingency. Its paradox is that even while it embraces uncertainty, it offers a safer home, existentially, than either community or universality. Both of those require suppression of individuals; their lived reality would be endangered. The aim of a relational narrative is to provide a place of shared and thus safer contingency. Through engagement patient and nurse are able to co-author an interpretation that may be more inhabitable than either of their personal narratives. Together, they can imagine a good in their situation that answers the ethical question of what should be done.

It is worth noting again that a dialectic without a *telos* can proceed indefinitely. Without closure, a framework such as I propose can give rise to further regions. Philosophy, in other words, need not culminate in the postmodern. The most interesting question provoked by a framework such as mine may be, "What now?" Treating the postmodern as the start of a new dialectic could be a project for further work. Indeed, given the popularity of the postmodern, a new dialectic already may be underway, with its first level of immediacy being uncritical immersion.

Logically (and ironically) my framework of ethical regions is itself postmodern only if I resist positing the postmodern as the last word, the high ground from which to gaze down upon other regions. Engagement, after all, may characterize a region some find inhospitable; their ethical homeland is communal or universal, not relational. Nursing can be enriched by that diversity, as long as we maintain a facility for appreciating and accommodating those who live in regions other than our own. If ethics is the alternative to force, as I believe, then we can try only to recruit — never require — others to move into our region. Meeting nonlocals on hospital ethics committees, for example, offers opportunities for recruitment; at a minimum it requires reconciling regional differences. (Perhaps that reconciliation constitutes a relational narrative.)

5. Where do you see the field of philosophy of nursing to be headed, including the prospects for progress regarding the issues you take to be most important?

I would like to see further philosophical development in at least two directions, epistemological and methodological.

1) Empowerment of patients seems increasingly to involve what could be termed information enchantment. Impersonal knowledge rendered in an apparently apolitical narrative via clinical websites may enchant nurse and patient alike to such an extent that no relational narrative seems needed. This may represent a new epistemological region only now accessible and perhaps safer than engagement. Relational ethics, after all, has risks: greater harm (as well as good) is possible, the closer individuals are to one another; exploitation and abuse can occur within intimate relationships. Perhaps a rationalist view of autonomy as needing only information, not assistance, is a safeguard against the hazards of engagement.

On the other hand, perhaps the new enchantment can be located within a dialectic of clinical assessment. Such a framework might identify a spectrum of epistemic approaches, including the immediacy of subjective symptoms, the objectivity of diagnostic categories, and the relational knowledge possible when patient and nurse combine those approaches. Viewed in this way, the enchantment with information may represent a welcome enhancement of the objectivity important at some point for both persons.

Traditionally, the translation of a patient's subjective experience into objective assessment has been the province of the professional. With access to virtually unlimited information, patients now may expect (or be expected) to make the translation themselves. If the process of clinical assessment is thought to comprise only this step, without attention to other kinds of knowledge, then patients will be considered sufficiently informed to make clinical decisions unassisted. This fits a rationalist approach, in which autonomy is merely the capacity for objective understanding. But it fails as an existential view of empowerment, which needs to include subjective and intersubjective as well as objective knowledge. Using a dialectical epistemology, each of these ways of knowing has a place, and the balance among them readjusts when one of them — in this case information — expands to the point of enchantment.

2) Another development in nursing philosophy hopefully will be an examination of philosophy's own methods. One that I find especially intriguing is language itself, a medium so unquestioned that we hardly recognize it as a method. We barely notice the disembodied theoretical

language with which we describe embodiment.

A theoretical framework that embraces experience in all its dimensions, especially its material reality, needs to recover the materiality of language. Philosophical text needs to express not only the weightlessness of concepts but the solidity of bodies, if it is to convey an understanding of experience. A postmodern philosophy that embraces the contingency of existence, its mutability and uncertainty, needs to recover the mercurial character of metaphor.

One avenue for the study of language as method is the contrast between literature and philosophy. While poetry dwells in particularity, philosophy aims for generality; poetry engages, philosophy reflects. Using language as metaphor, poetry makes the world unfamiliar, surprising, while philosophy aims for certainty. Philosophy shares with poetry the attempt to offer a new understanding of the world by redescribing it, but unlike poets, philosophers tend to view each new description as final; we want our surprises to become true.

Philosophy can be saved from that finality by learning to use language as poets do – not by writing verse but by sharing the poet's project of making contingent what had become literal, then making the redescription contingent in turn, before it sounds true. That rhythm of assertion and negation, of certainty then surprise, is precisely the meter of dialectic, a method that constructs endless new contingencies. Language that assures and undermines at the same time, that asserts as well as refuses certainty, is ironic. Philosophers would become ironists.

What would this view of language mean for philosophy? It would undermine the exclusive authority of the philosopher's voice, admitting the validity of other voices. Further, it would compel us to invent new textual forms in which language retains its sensual character. This would be even more challenging than philosophers composing poetry, because it would mean combining the arid heights of abstract thought with the vivid imagery that metaphor creates.

Texts like that could be unsettling. They might resemble, for example, a motet, in which competing and contradictory voices combine, each with its own perspective, without any of them attaining dominance. Voices would interrupt as well as enhance each other, driving to distraction a reader looking for linear logic. A multivocal text could complement my own language as a means of disrupting the orderly progression of dialectic. Instead of clearly defined levels, a motet would represent overlapping areas with unclear edges, a tangle of voices vying with each other, not even my voice having the last word.

Selected Works

Gadow, Sally:

1980. "Body and Self: A Dialectic." *Journal of Medicine and Philosophy* 5: 172-185.

1980. "Existential Advocacy: Philosophical Foundation of Nursing." In *Nursing: Images and Ideals; Opening Dialogue with the Humanities*, edited by Stuart Spicker and Sally Gadow, 79-101. New York: Springer Publishing Company.

1983. "Frailty and Strength: The Dialectic in Aging." *Gerontologist* 23: 144-147.

1988. "Remembered in the Body: Pain and Moral Uncertainty." In *Dax's Case: Essays in Medical Ethics and Human Meaning*, edited by Lonnie Kliever, 151-167. Dallas, TX: Southern Methodist University Press.

1992. "Existential Ecology: The Human/Natural World." *Social Science and Medicine* 35: 597-602.

1994. "Whose Body? Whose Story? Questions about Narrative in Women's Health Care." *Soundings* 77: 295-307.

1995. "Narrative and Exploration: Toward a Poetics of Knowledge in Nursing." *Nursing Inquiry* 2: 211-214.

1995. "Clinical Epistemology: A Dialectic of Nursing Assessment." *Canadian Journal of Nursing Research* 27: 25-34.

1999. "Relational Narrative: The Postmodern Turn in Nursing Ethics." *Scholarly Inquiry for Nursing Practice* 13: 57-69.

2000. "I Felt an Island Rising: Interpretive Inquiry as Motet." *Nursing Inquiry* 7: 209-214.

2000. "Philosophy as Falling: Aiming for Grace." *Nursing Philosophy* 1: 89-97.

2003. "Restorative Nursing: Toward a Philosophy of Postmodern Punishment." *Nursing Philosophy* 4: 161-167.

"Sally Gadow" http://www. sallygadow. com

8

Ann Gallagher

Reader in Nursing Ethics

Faculty of Health and Medical Sciences, University of Surrey, UK

---

1. How were you initially drawn to philosophical issues regarding nursing?

At the age of seven I was led to believe that I was responsible for my 'sins' and that if I repented I would be forgiven. I went to confession regularly and shared my list of transgressions of the Ten Commandments with a priest. These may have ranged from telling lies to being disobedient and having dishonourable thoughts about a parent. Sometimes I struggled to come up with material for the monthly unburdening. I would then worry that I had to have something to confess but feared that this would have to involve some fabrication which could be construed as lying. Already I had a sense of the complexity and ambiguity of the moral life.

When I was twelve years old, thirteen people were shot dead by soldiers 10 miles from my home. This occurred in Derry, Northern Ireland, during a civil rights march and became known as "Bloody Sunday". I was a student nurse at the Royal Victoria Hospital in Belfast in 1981 when ten young men died on hunger strike. We were advised that we would have to report to the hospital to help out should 'all hell break loose' when the first hunger striker died. The men were republican prisoners who were protesting because they wanted to be treated as "prisoners of war" rather than as ordinary criminals. These were not the only events that impacted on the people of Northern Ireland or on the health services during what became known as "the Troubles". Bombings and shootings were commonplace and over 3, 500 people died from violence during this period.

As student nurses, we provided care to many civilians - most of those hospitalised were being treated for regular illnesses - and to British soldiers, those labelled "paramilitary" and others who had been injured during this time. I recall a number of patients admitted for surgery following "kneecappings" or "punishment shootings". These were mostly young men who had allegedly engaged in activities such as drug-dealing, considered punishable by paramilitary groups. One experience that I remember vividly was caring for a young woman of my age who had been shot coming from a protestant church. We communicated well, despite her tracheostomy, and she told me that she hoped to work in the probation services. When I was on days off, I heard on the news that she had died. Her hopes, and those of her family, were unrealised just because she was someone of a certain religion.

What I recall most acutely during this period was the expectation from others and ourselves that, as nurses, we would treat everyone the same and just get on with care delivery. I do not recall receiving any input on professional ethics. Unsurprisingly, around this time I was questioning religious perspectives, considering the possibility of free will in such challenging political and social environments and seeking some alternatives to the oppressive and sectarian model with which I had grown up. The possibility of a secular and humanistic foundation for values was compelling.

During my student nurse training I completed a placement in a large psychiatric hospital. I worked with a very supportive and impressive clinical nurse specialist who took every opportunity to guide students as they negotiated the uncertainties and challenges of acute psychiatric care. As a result of this placement I decided to apply for mental health training. I went to England to do this and became a Registered Mental Nurse. Later I worked as a charge nurse on a unit for adolescents with a range of mental health problems. One of the social workers who led the daily community meeting used metaphors to capture the mood of the group. I discovered that he had read for a philosophy degree before becoming a social worker.

In my late twenties, I was fortunate to receive funding to read for a combined degree in philosophy and health studies, a degree that straddled the humanities and social sciences. I had the opportunity to take courses in: American philosophy, the philosophy of psychoanalysis, existentialism, moral philosophy and social theory and policy. My degree programme coincided with the premiership of Margaret Thatcher and the introduction in the UK of the unpopular poll tax. My university had an active revolutionary communist group and they voted to occupy the university in protest at the poll tax. On the day of the vote we had an outside speaker on the existentialist theme "Man is condemned to be free". Outside the lecture theatre, we could hear the clanking of chains as students and staff were locked in (and locked out). The paradox of the situation and the challenge of political action that conflicted with self interest (I wanted to attend my lectures, use the library etc) were both ironic and frustrating.

I then went on to do a Masters' programme in medical and social ethics and a PhD in professional ethics. Along the way, in addition to my studies, I was influenced by reading and theatre. A reading of *The Women's Room* by Marilyn French (1977) most likely stoked my early feminist inclinations. Reading also the novels and work of Iris Murdoch was instructive. *The Sovereignty of Good* (Murdoch 1971) remains a source of moral guidance as I strive to counter my inclinations to be overly and uncharitably critical. As Murdoch puts it: "the idea of a just and loving gaze directed upon an individual reality. I believe this to be the characteristic and proper mark of the active moral agent" (33).

Visits to the theatre in Belfast, Edinburgh and later in London made me aware of the potential of the arts to interrogate contemporary ethical issues and to challenge the audience. Most influential were theatre performances in the 1980's and '90's of *Juno and the Paycock* (O'Casey 1924), *One Flew Over the Cuckoo's Nest* (Kesey 1962), *Whose life is it anyway?* (Clark 1972), *Torch Song Trilogy* (Fierstein 1982) and *Oleanna* (Mamet 1994).

As I was beginning my PhD programme I worked as a tutor at Open University summer schools. It was there that I was introduced to 'virtue ethics' and the work of philosopher Rozalind Hursthouse. Here was a moral theory, I thought, with the potential to relate meaningfully to healthcare practice. Her book *Beginning Lives* (1987) served as my first reading in this area of moral philosophy. Virtue ethics became the theme of my PhD study *Healthcare Virtues and Professional Education* (2003). My PhD supervisor, Professor Ruth Chadwick, was also very influential enabling me to better appreciate the value of philosophy in relation to everyday healthcare practice and indeed the value of healthcare practitioners engaging in philosophy. In response to a discussion about the relationship between ethics and professional practice, Ruth recommended that I read Warnock's (1971) *The Object of Morality*. The idea that humans require morality because we are limited in rationality, sympathy and in competition for resources remains a very plausible account and one that backgrounds my thinking about healthcare ethics. Warnock suggests that:

> The 'general object' of morality, appreciation of which may enable us to understand the basis of moral evaluation, to contribute to betterment – or non-deterioration – of the human predicament, primarily and essentially by seeking to countervail 'limited sympathies' and their potentially most damaging effects. (26)

My response to the question 'how were you initially drawn to philosophical issues regarding nursing?' may be viewed as a rather circuitous ramble through thorny political and professional terrain. What became clearer to me are the inextricable connections amongst the personal, political, professional and philosophical. My interest in philosophical issues in nursing was preceded by a broader interest in philosophical questions, relating to free will and the practical and philosophical foundations for the moral life. These questions remain pertinent and inform my work in nursing ethics.

2. What, in your view, are the most interesting, important, or pressing problems in contemporary philosophy of nursing?

I recall seeing a cartoon of a rather beleaguered nurse arriving at the top of a mountain to seek the advice of an elderly loin clothed sage. The nurse asks 'what is the meaning of nursing?' This is an interesting philosophical question and there is no doubt that nurse philosophers will continue to try and pin down the essence of a very diverse practice.

In my view the most important or pressing problem we need to engage with relates to how we understand and sustain ethical care practices and respond to unethical care practices. There have been many reports of deficits in care in the UK (and in other countries) whereby patients, often the most vulnerable, experience abuse and neglect in care services. Some of the reports have attributed care deficits to 'bad, cruel nurses' (Patients Association 2009), whilst others have acknowledged the importance of considering micro (individual); meso (organisational); and macro (social/political) perspectives (Royal College of Nursing 2009).

Reports of unethical practice in hospitals and institutions are not new. In the 1967 report *sans everything: a case to answer,* The author, Barbara Robb, brings together many harrowing accounts of the inhumane treatment of older patients in so-called care institutions at that time. It is sometimes forgotten that nurses and nurse leaders in earlier eras grappled with philosophical and practical aspects of care deficits, that they suggested solutions and referred to similar ethical concepts as we would today. In Robb's report, nurse W. J. A. Kirkpatrick (Robb 1967, 56) stated:

> Nurses are often praised for their strength in helping other people. Needed now is the courage of truth to help themselves. Nurses must be for nursing, supporting its advances in every conceivable manner. If they do this, they will uphold the right of the sick and troubled in mind to be regarded as

> members of the human race; they will uphold the dignity of their patients, their own nursing profession and the National Health Service. For nursing is a career of compassion towards men and women by men and women, in which the art of nursing becomes and is an act of love towards the patient and vice versa, and between nurse and fellow-nurse.

Kirkpatrick (Robb 1967 p. 57) suggests a definition of love as 'the accurate estimate and supply of someone else's need' and concludes that we need to remember that 'in all ages nurses have been, as they still are, consecrated (at their own desire) to relieve suffering and *to care always*. Do we?'

What recent reports relating to nursing practice suggest is that nurses do not *care always*. It is a matter of concern - described by some others as a 'crisis' in care or a 'crisis of compassion' - that there is such relentless reporting of unethical care practices in 2012. Ethics is now a compulsory component of the professional curriculum and there is much research and scholarly activity in philosophical and empirical ethics.

Why then, is there such an apparent disconnect between what is taught and studied and what occurs in practice settings? How should we think about this? Indeed, what should be taught and studied so that the values of nursing are transmitted effectively and applied successfully to enhance the flourishing of patients, families and practitioners? What research paradigms are appropriate for empirical nursing ethics? Crucially, how can the articulation of the value and values of nursing contribute to the development of sustainable ethical practice? Can we agree on values that are transcultural, crossing country and continent boundaries?

These are some of the most pressing areas of concern in contemporary nursing philosophy

3. What, if any, practical and/or socio-political obligations follow from studying nursing from a philosophical perspective?

The field of nursing ethics offers a wide range of explanations for, and potential solutions to, the challenges outlined above. Philosophical scholarship published in *Nursing Ethics* and other journals and textbooks contribute to the development of our understanding of concepts such as dignity, respect, moral sensitivity, consent, compassion and futility and to the development of theories such as virtue ethics, care ethics and principles-based approaches. Empirical research in nursing ethics offers a wide range of perspectives and insights regarding the factors that contribute to ethical and unethical practice. Research, for example, relating to moral distress and the ethical climate of organisations suggest a very plausible relationship between the two (See for example, Corley at al 2005). Empirical work relating to dignity in care, ethics education and research ethics also suggests how practice might be improved in healthcare, educational and research contexts.

Regarding the practical and/or socio-political obligations that follow from studying nursing from the perspective of philosophical and empirical ethics, these appear to be:

The obligation to be skeptical and reflective in interrogating concepts, practice and policies asking "are these ideas or arguments, practices and policies in accord with the ethical foundations of nursing?";

The obligation to engage with, and disseminate, published scholarship and research in nursing ethics asking "how does this relate to, and how might it develop, my practice?";

The obligation to contextualize an understanding of ethical and unethical practices enquiring about the inter-relationship amongst individual, organizational and social, political and global factors drawing on inter-disciplinary perspectives;

The obligation to take the time to recognize and celebrate ethical care practices and to be courageous enough to respond effectively to unethical practices so that the wellbeing of patients and practitioners is safeguarded; and

The obligation not to be complacent recognizing our own fallibility and vulnerability and aspiring to betterment in our personal and professional lives.

4. In what ways does your work seek to contribute to philosophy of nursing?

The focus of my work relates to an understanding of the philosophical and empirical aspects of ethical and unethical care practices, most particularly those engaged in by nurses. I seek also to understand what underlies ethical care practices and what contributes to unethical practices. Much of my early work in healthcare ethics involved writing and teaching aspects of philosophical ethics, for example, in relation to end of life care, medical and nursing ethics and human rights (see selected publications below). My initiation into empirical ethics occurred when I worked as research assistant with the philosopher, Professor David Seedhouse. Professor Seedhouse had completed a book relating to key concepts in nursing philosophy and wished to design an empirical project in response to the question 'Does teaching nurses about (key concept in nursing philosophy) improve practice?' We agreed to focus on 'dignity' and designed a quasi-experiment to respond to this question. This was both a challenging and a very rewarding project conducted on three sites over one year. I rapidly became aware of the limitations of a positivist paradigm in healthcare research and the value of qualitative work.

We published the findings in professional and peer-reviewed journals. This signalled my engagement with the concept of 'dignity' and also 'respect' and I published in journals and books with a view to clarifying these concepts for healthcare students and practitioners (See selected publications below). I had the opportunity to develop empirical work when I was commissioned - with the late Professor Paul Wainwright and Dr Lesley Baillie - by the UK Royal College of Nursing to conduct a survey of nurses' perspectives on dignity in care. We analysed over 2000 responses and the resulting report *Defending Dignity: Challenges and opportunities for nursing* has been used to underpin the development of teaching materials and is regularly cited in dignity reports.

In other research I collaborated with colleagues to seek the perspectives of service users regarding their experience of receiving good and bad news about their mental health. I also was a team member on a research study exploring how general practitioners (family physicians) negotiated conflicts of interests in safeguarding children. I co-led the Delphi process of this study with Professor Wainwright. The relationship between philosophical and empirical work in applied ethics continues to be of great interest to me.

Ethics is sometimes perceived to be a rather esoteric and inaccessible discipline. In my teaching and speaking activities I aim, therefore, to make this area of academic study not just accessible but also engaging, meaningful and helpful to students. Ethics enables practitioners to identify the key ethical aspects of a practice situation, to interrogate and apply key ethical concepts that illuminate the situation and facilitates decision-making and practice supportive of professional accountability. It is hoped that engagement with ethics will lead students and practitioners to develop necessary knowledge to underpin their professional accountability and skills to practice ethically. Along the way, it is hoped they continue to develop healthcare virtues such as respectfulness, courage, wisdom, justice and integrity.

In the role of Editor of the international journal *Nursing Ethics* I am privileged to work with Editorial Board members and others who so generously share their expertise and experience in the field. Together we aim to disseminate the findings of philosophical scholarship and empirical research that illuminates ethical and unethical care practices. It is a privilege to receive manuscripts from colleagues from all parts of the globe, manuscripts that provide opportunities for us to learn from each other and to develop transcultural perspectives. Engagement with colleagues who work in the field of nursing philosophy and with publications in the journal *Nursing Philosophy* also facilitate a broader view of the relationship between philosophical work and practice perspectives on care.

5. Where do you see the field of philosophy of nursing to be headed, including the prospects for progress regarding the issues you take to be most important?

This is a very exciting and also a challenging time for care professions. There is escalating care need. Many people are succumbing to dementia and as people live longer there is also more chronic disease and more complex care requirements. Alongside this there are efforts to contain costs, to manage scarce healthcare resources and to attract and retain healthcare practitioners who not only have highly developed knowledge and technical skills but also have the necessary moral qualities. The brightest and the best of our young people have many more choices regarding education and career and they may not opt for a profession that appears to require resilience and tolerance of poor conditions and lower salaries than other professions.

The direction of travel of nursing philosophy overall is impossible to predict, however, there are observable trends within nursing ethics. There is, for example, an increased interest in qualitative perspectives on ethical aspects of healthcare practice from the perspectives of patients, families, students and practitioners. There are increased efforts to measure ethical aspects of practice (for example, moral distress, moral sensitivity, moral judgement and ethical climate). There is also sustained engagement with 'what works' in relation to ethical practice, for example, research that evaluates innovative educational and practice initiatives designed to promote ethical practice. All of this is to be welcomed. However, it is not enough.

We need to continue to develop philosophical perspectives that illuminate the nature of ethical healthcare practice. We need also to be ever-ready to critically discuss the meaning of concepts such as care, dignity, compassion and integrity. We need to consider how a particular action or intervention will benefit patients, families and/or practitioners. And ask: What harm or distress might ensue from this intervention? What impact might the intervention have on the dignity of the person? Is the proposed action fair or just, that is, does it include or exclude the most marginalised or disadvantaged in our society? In our everyday work as nurses, teachers, researchers and managers we need to consider and demonstrate how philosophical concepts can contribute to our articulation of the value and values of nursing so all can hear (most particularly the next generation of nurses).

Crucially, we also need to overcome any inclination we might have to be complacent. However illuminative the philosophical and empirical perspectives proposed by our disciplines, we need to accept that we will not have all the answers to the complex challenges that confront us now and in the future. In relation to the challenge of unethical care practices, it seems clear that nursing philosophy offers no panacea and nurse philosophers cannot go it alone. My proposal therefore is that we look outwards to other disciplines, such as social and organisational psychology, the arts and perhaps also neuroscience, for alternative perspectives. In inter-disciplinary discourse we need to articulate loudly and clearly the contribution of nursing ethics and nursing philosophy can make to other disciplines and professions.

References

Hursthouse R. 1987. *Beginning Lives*. Oxford: Blackwell in Association with the Open University.

Murdoch I. 1970. *The Sovereignty of Good*. London: Routledge & Kegan-Paul.

Robb B. 1967. *sans everything: a case to answer* London: Thomas Nelson & Sons Ltd.

Warnock G. J. 1971. *The Object of Morality* London: Metheun & Co Ltd.

Selected Works

Books

Chadwick R. W. Tadd and A. Gallagher. (in preparation - due for publication 2013) *Ethics & Nursing Practice: A Case Study Approach* (2nd ed). Palgrave/MacMillan

Gallagher A. and S. Hodge. eds. 2012. *Ethical, Legal and Professional Aspects in Healthcare: A Practice-based Approach*. Palgrave/ MacMillan

Banks S. and A. Gallagher. 2009. *Ethics in Professional Life: Virtues for Health and Social Care*. Palgrave/MacMillan.

McHale J. and A. Gallagher. 2003. *Human Rights and Nursing*. Butterworth Heinemann/Elsevier.

Articles & Reports

Gallagher A., P. Wainwright, H. Tompsett and C. Atkins. 2012. "Findings from a Delphi Exercise Regarding Conflicts of Interests, General Practitioners and Safeguarding Children: 'Listen carefully, judge slowly'." *Journal of Medical Ethics* 38: 87-92

Gallagher A. and C. Gannon. 2011. "Difficult Decisions in Cancer Care Conducting an Ethics Case Analysis." *European Oncology and Haematology* 7: 101-105.

Gallagher A. 2011. "Ethical Issues in Patient Restraint." *Nursing Times* 107: 18-20

Gallagher A. 2010. "Whistleblowing: What Influences Nurses on Whether to Report Poor Practice?" *Nursing Times* 106: 22-25

Gallagher A., A. Arber, R. Chaplin and A. Quirk. 2010. "Service Users' Experience of Receiving Bad News About their Mental Health." *Journal of Mental Health* 19: 34-41.

Gallagher A. and V. Tschudin V. 2010. "Education for Ethical Leadership." *Nurse Education Today* 30: 224-227.

Gallagher A., K. Horton, V. Tschudin and S. Lister. 2009. "Exploring the Views of Patients with Cancer on What Makes a Good Nurse – A Pilot Study." *Nursing Times* 105: 24-27.

Baillie L., A. Gallagher and P. Wainwright P. 2008. *Defending Dignity: Challenges and Opportunities for Nursing*. Royal College of Nursing, London [Research report - http://www. rcn. org. uk/__data/ assets/pdf_file/0011/166655/003257. pdf}

Gallagher A., S. Li, P. Wainwright, I. Rees Jones and D. Lee. 2008. "Dignity in the Care of Older People - A Review of the Theoretical and Empirical Literature." *BMC Nursing* 7: 11. http://www. biomedcentral. com/1472-6955/7/11

Wainwright P. and A. Gallagher. 2008. "On Different Types of Dignity in Nursing Care: A Critique of Nordenfelt." *Nursing Philosophy* 9: 46-54.

Gallagher A. 2007. "The Respectful Nurse" *Nursing Ethics* 14: 360-371.

Gallagher A. and P. Wainwright. 2007. "Terminal Sedation: Promoting Ethical Nursing Practice." *Nursing Standard* 21: 42-46

Wainwright P. and A. Gallagher. 2007. "Ethical Aspects of Withdrawing and Withholding Treatment." *Nursing Standard* 21: 46-50.

Gallagher A. 2004. "Dignity and Respect for Dignity – Two Key Health

Professional Values: Implications for Everyday Nursing Practice." *Nursing Ethics* 11: 587-599.

Seedhouse D. and A. Gallagher. 2002. "Undignifying Institutions." *Journal of Medical Ethics* 28: 368-372.

Gallagher A. 1995. "Medical and Nursing Ethics; Never the Twain?" *Nursing Ethics* 2 (2)

Media

Gallagher A. 2012. "The key to improving nursing lies with nurses themselves." *The Times* 04/05/12 p. 61 (Public Sector section)

Gallagher A. 2012. "How nursing should tackle its image problem." *The Guardian* 10/05/2012
(See http://www. guardian. co. uk/healthcare-network/2012/may/10/ nursing-should-tackle-image-problem#start-of-comments)

Gallagher A. (2013) The ethics of force-feeding inmates: room for debate, *New York Times* 1st May
(see http://www. nytimes. com/roomfordebate/2013/05/01/the-ethics -of-force-feeding-inmates)

Gallagher A. (2013) Care: A higher calling Times Higher Education 28th February
(see http://www. timeshighereducation. co. uk/comment/opinion/ care-a-higher-calling/2002098. article)

Gallagher A. (2013) Slow ethics will tackle moral winter *Times Higher Education* 10th January (see http://www. timeshighereducation. co. uk/ 422322. article)

9

Dave Holmes

Professor, School of Nursing, Faculty of Health Sciences
University of Ottawa, Ottawa, Canada

---

1. How were you initially drawn to philosophical issues regarding nursing?

In 1997, I began my master's degree at the University of Montreal in the Faculty of Nursing, and then fast-tracked into the PhD program there. Through both degrees I had two supervisors, one a sociologist and the other a criminologist, both of whom did critical work in nursing and health. The fact that they were not nurses was a concern for many professors working at the Faculty. The summer before starting my program, both supervisors insisted that I read Michel Foucault's (1975) *Discipline and Punish*. They requested this because, at the time, I was a nurse working in a maximum security forensic psychiatric setting in Montreal. I rapidly understood that I had no choice but to read that book. As a consequence, I spent most days of the week, that summer, at *Parc Lafontaine* in Montreal having my first experience in (political) philosophy. I remember being quite distressed when reading Foucault's book, as I became suddenly very critical of nursing practices in forensic psychiatric settings. Nursing practice was contaminated by something that had more to do with power and social control than anything else. What was happening at the hospital/prison was exactly what I was reading about in Foucault's book. I started analysing my own practice and questioned what I was doing.

Knowing full well that I would not understand everything at first, my supervisors then requested that I read Foucault's (1976) *History of Sexuality*, where I would learn about the concept of biopower. It helped me to put the concepts that I had learned in the previous book into the context of public health and psychiatry, my two fields of professional practice. That summer transformed my thinking; from this introduction to political philosophy, I began to read poststructuralist and other critical literature. Although Goffman had more to do with sociology and anthropology, I tried to merge Foucault and Goffman in a rigorous manner, and then was exposed to Derrida, and finally Deleuze and Guattari. I continued exploring these difficult texts during my postdoctoral experience, where again I had two supervisors, one in Social Work and the other in Public Health. My main supervisor said something that I have never forgotten: "Listen to me carefully, Dave. Research designs and methods – forget that. That is easy stuff. Get into theory and philosophy. I want you to expand your understanding of Derrida, Deleuze and Guattari and others (such as Kristeva and Virilio) as much as you did for Foucault. Push your brain a bit more – it's more difficult but you'll find it worthwhile." So I began to apply Deleuze and Guattari to public health, because while I continued to be a full-time psychiatric nurse I was working as a part-time street nurse in Montreal, caring for prostitutes, homeless persons, and many drug users. When I came into contact with men having unsafe anal sex, Professor Chambon said to me, "Look deeper into Deleuze and Guattari."

It was my supervisors (PhD and postdoctoral), whom I consider to be influences from outside nursing, who encouraged me to consider the field of philosophy as a whole, and to look at other critical authors as well. As soon as I had touched and tasted this new body of literature, I knew that I could never return to those boring and inarticulate nursing conceptual models.

Everything that I'm doing in nursing at the moment takes a philosophical and political tone. Even with some topics like evidence-based nursing, I can never just look at it from an empirical perspective. Also, the book that I co-edited two years ago on ethics questioned the concept of autonomy, and again it was influenced by Foucault and the way people are subjected to some kind of power and control (Rudge and Holmes 2010). I've also been using other concepts lately, like abjection from Julia Kristeva – more of a psychoanalytic field but since I'm a psychiatric nurse it was not a difficult apprenticeship – and using Žižek to look at violence (Holmes, Rudge and Perron 2012). I cannot escape it – I cannot supervise without it and I cannot think outside of it. I do, however, consider myself a conservative researcher in the sense that I continue to go back to Foucault, Derrida, Goffman, etc., and do not use fashionable authors. I'm always trying to refine my own thinking going back to these basic/classic texts.

But, there are also great personal challenges, even risks, in using these texts. Some are in the actual writing – because of its conceptual nature I have to begin with pen and paper, and in French, to wrestle with the concepts, to put them together before I can move to the computer. Other risks derive from the ideas that you come up with. I remember when I was looking at unsafe anal sex between men, the empirical data led me to look at Georges Bataille's work on "limit experience". To read Georges Bataille's *Eroticism* was scary because it is about transgression of whatever is sacred, and this in turn led me to go back to the Marquis de Sade. When I asked a colleague to help me with one particular interview, using these texts as a way to understand it, he declined, saying it was too much and he could not engage in that. So I had to destroy the interview – I had touched a spot where I could not navigate – where I could not cope, not because I could not understand Bataille's work but because it provided an understanding of a radical sexual practice that I could not engage with.

2. What in your view are the most interesting, important, or pressing problems in contemporary philosophy of nursing?

I think one of the most difficult problems, but likely the most important one, is to be able to teach it. To be honest, I don't think we have nursing philosophers. I believe that if you use phenomenology, for example, you should go back to Heidegger, and if you want to understand poststructuralism, you need to read Derrida, Deleuze, and Foucault. They were not nurses but you need to bring these philosophers and their concepts into nursing and teach the basics of their works. I think that people are afraid, perhaps because the concepts they use may seem abstract and sometimes not very accessible. I teach undergraduate and graduate students and I always try to define the concept I'm using – sort of a concept analysis – and then apply it to the empirical work that nurses do. For example, I've used the concept of abjection to look at semen exchange; I've used the concept of abjection to look at HIV contamination. Abjection is a highly philosophical concept and a difficult one to understand but you have to read and re-read – you just don't grasp Kristeva's ideas in a moment. Deleuze's work, too, can be maddening. But when you get even a basic understanding of these concepts, they illuminate your work.

The pressing thing to me is to open this route in nursing and to show nurses and the nursing community that what we experience on a daily basis can be explained by theories outside the ones we are often "forced" to use in nursing. To me it's an exercise – a challenge that admittedly comes easier now than it did in the beginning – to apply these concepts to something concrete in order to make sense of what is going on in clinical settings. Now, for example, I understand power from different perspectives – from Marx, from Foucault, from Weber – and I'm able to play with them a bit more. At one point we were engaged in Canadian Institute for Health Research funded research in three bathhouses, looking at unsafe sex between men. So why in nursing would we have something to say about a penis, a wall, a hole and a mouth – all parts of a dark room where one can find glory holes? We bothered because we had something to say from an ethical perspective from which we could critique public health discourses regarding anonymous sex. Using Deleuze and Guattari's concept of *machinic assemblages*, we understood the link between these elements in such a way that we were able to offer another type of explanations for anonymous sex beyond the routine public health warnings about the dangers of syphilis, etc. (Holmes, O'Byrne and Murray 2010). Publishing the resulting paper was somewhat difficult – although the editor agreed that it was well written and grounded philosophically, he was uncomfortable with the graphic nature of the topic. But it was finally published after being carefully reviewed. As academics, we have the duty to expose nurses to these concepts and challenge them to look at their practice in light of these concepts. In so doing, we engage in a pedagogico-political action that pushes the boundaries of our discipline that is so conservative. Students and nurses need to know that there are opportunities to think differently.

3. What, if any, practical and/or socio-political obligations follow from studying nursing from a philosophical perspective?

I do not believe that you can study philosophy in nursing without being political. In the past, I have looked at capital punishment in the US, where nurses are part of the trained healthcare personnel delivering lethal injections, and argued that this activity not only conflicted with nursing ethics but also allowed for the skilled knowledge and work of nurses to be appropriated by the state penal machinery (Holmes and Federman 2003). Much the same thing was happening with nurses and the CIA in Guantanamo, which we also wrote about (Federman and Holmes 2011). We must continue to expose these practices, not in a journalistic way but from ethical and political perspectives, to show shifts in the discourse. As nurses, what are we doing in prisons or on the street with prostitutes and drug users when we report them to the police? Who is the client? What are empowerment and autonomy, and are they truly possible when there is an imbalance of power between individuals?

The last research I conducted in a prison setting resulted in a paper entitled "Civilizing the 'Barbarian'." Here we claimed that "behaviour modification programs violate[d] both scientific and ethical norms in the name of doing 'what is best' for the patients," and we contended that their continued use constitutes an unethical approach to the nursing care of mentally ill offenders (Holmes and Murray 2011). I knew that publishing this paper would mean I would not be welcome in that setting again, but I also knew that I had a duty to say something. We used concepts from Goffman, Deleuze and Guattari, and Foucault, to show how unethical these practices were – and that the barbarian was not the patient. The implications for publishing that piece were very serious.

Socially and politically I felt obligated to say something. I could have just hidden behind the data, written two or three innocuous papers, and continued my research, because I had been funded to undertake another project in a nearby unit. But I did not do that, and this paper caused the major controversy of my career. I don't consider the paper offensive, because otherwise it would not have been published, but it is blunt and direct. We criticize the programs, discuss the impossibility of autonomy, and describe the Foucauldian concept of surveillance using the metaphor of the panopticon. How dare we ask patients to be part of this kind of madness? A National Post journalist found the paper and contacted a psychiatrist involved, who not only accused us of having no ethics approval for the research but also insinuated that the paper was a false representation of the reality. In fact, we had had two sets of ethics approval and the research funds had been released. The journal editor stood behind our paper and invited the psychiatrist, who had made a formal complaint to the journal, to provide a response, but they did not really engage with the arguments and in the end, had nothing worthwhile to say. Nurses were opposed to the practice in vogue in the setting and our research aimed at understanding nurses' experience in a forensic psychiatric setting. I am a full professor and at the time was Vice-Dean, Academic, so my career was not affected, but the result might have been different if this had happened to someone who was just starting out his or her academic career. Nevertheless, the whole episode was both distressful and stressful to me – but again, that's the risk. Would I do it again, to fight censorship of research: no doubt that I would. I wrote with colleagues about this experience (Perron, Jacobs and Holmes forthcoming).

What this issue has pointed out to me even more strongly, however, is that psychiatry is an apparatus – a *dispositif* with discourses and practices intertwined. My next book will be a collection of eighteen other like-minded authors who are exposing this psychiatric apparatus. Using the ideas of Foucauldian power and *governmentality*, they will be criticizing, for example, the Diagnostic and Statistical Manual of Mental Disorders (DSM) 3 and 4, or the link between psychiatry and pharmaceuticals. Thomas Szasz, before his death in September 2012, agreed to open the book with a very thoughtful chapter on the history of psychiatry. Of course, I'm not saying that everything in psychiatry is bad because I do know that some people need help, and I do know that some people need medication – that's not my point. The point is to look at how psychiatry can be excessive and in some settings can be very brutal both against patients and nurses. I continue to construct my work using these concepts and I think I have a duty as a researcher to expose these types of structures even at the risk of being attacked or discredited or ridiculed. I have a political purpose and that is to change things.

4. In what ways does your work seek to contribute to philosophy of nursing?

I'm using philosophies from outside nursing and then applying them to nursing experiences or domains. I think that the concepts I'm using make sense in unpacking, explaining and critiquing some nursing and institutional practices. I'd like nurses to explore their own practices; I'd like them to think about what they do. I'm not interested in following trends – and perhaps this is a bit of a critique of my colleagues as well as an indication that I'm a little conservative – but I always go back to Foucault, whether or not he's still in vogue in nursing. But I'd like my work to at least be connected to my personal and professional experiences through the way I think about nursing. And I want both undergraduate and graduate students to read books or papers by these philosophers or their followers and to finish them feeling a kind of malaise, to be a bit unsettled. I'd like them to feel like this so they can begin to think for themselves. Most of all, I like to think that the work I'm doing creates movement.

I'm not saying that I'm always right because some people responding to my work don't always agree with what I've written, but at least they've engaged with the subject. And I'm not doing this for myself or otherwise I would not write. I admit it does take courage to write about glory holes or semen exchange and then critique public health authorities. And when I speak about evidence-based nursing as a fascist way of developing knowledge, I'm not using the concept just to shock but to point out that if you are focused solely on using evidence-based ideology, then you are excluding other ways of developing knowledge and that this is a fascist-type ideology (Holmes et al. 2006). Of course we were criticized for this assessment but, for example, veterinarians from as far away as the University of Zurich thanked us for our paper and said that finally someone was critiquing this trend; we received the same comment coming from Harvard's School of Medicine.

5. Where do you see the field of philosophy of nursing to be headed, including the prospects for progress regarding the issues you take to be most important?

I think that the most important issue is to make nursing philosophy accessible. Although I know it seems highly abstract, it has to be understandable, applied. And I think that our job is to make our papers understandable to many different kinds of people, so that these concepts can be taken up and applied, brought into practice. I once received an email from an RN who wrote that she was sitting out in a bush in Australia and had just finished my paper, "The Rise of the Void," (Holmes et al.

2008) and she said, "thank you so much, you just made my day." I care about this woman out in the bush reading my texts in Australia taking the time to send me an email. And I care about the guy in San Francisco sending me a long letter telling me that he's read my work on *barebacking* (unsafe anal sex between men) where I used the idea of limit experience from Bataille (Holmes, O'Byrne and Gastaldo 2006). "Thank you so much for not 'pathologizing' us – thank you for understanding our practice the way that we feel it," he wrote. Of course from a public health perspective these activities cannot be good but it was not my job to condemn. The main objective of this research was to look at the practice from the perspective of the people doing it and to report on it and try to explain it. The question had been, why do these men exchange semen? I knew that from a political perspective I could use Deleuze's concept of resistance to suggest that it might be an act of resistance against normative discourses. Still, there seemed to be something even more symbolic with this kind of exchange so I worked with a psychoanalyst (Lacanian tradition) and came up with a paper, published in *Nursing Inquiry* years ago, on the anatomy of forbidden desire (Holmes and Warner 2005). This paper, which was daring at the time, resulted in a woman doctor telling me that, while she was too embarrassed to talk about the paper, she thought it was exceptional. It is responses like this that make me realize the importance of making your work accessible.

My primary hope is that nursing will become a bit more philosophical and ultimately more political. I don't believe that you need to be a *barebacker* to talk about *barebacking*, a drug user to talk about crack pipes, and you don't need to be a mentally ill person yourself to critique psychiatry; you just need to be sensitive to others and understand from their perspective where they are coming from. Then use the appropriate tools, which are for me philosophical ones, as weapons to deconstruct taken-for-granted nursing discourses and practices. I hope that nurses will engage more with philosophical texts, maybe not so much to enlighten their practice but to criticize it – to look closely at it and try to change it. That is my objective and what is most important to me. They have to be able to read these texts and we have to be able to provide them with the tools necessary for them to access these works and we have a duty as academics to be as crystal clear as we can. Even if the concepts can be very abstract or complex, we need to present them in a way that is so clear that they see a link with this and their practice. We will get nowhere if what we say does not make any sense and if what we say does not touch nurses in some way.

References

Federman, C. and D. Holmes. 2011. "Guantánamo Bodies: Law, Medicine and the Media." *Mediatropes* 3: 58-88.

Foucault, M. 1975. *Surveiller et punir: naissance de la prison*. Paris: Éditions Gallimard.

Foucault, M. 1976. *Histoire de la sexualité*. Paris: Gallimard.

Holmes, D. and C. Federman. 2003. "Killing for the State: The Darkest Side of American Nursing." *Nursing Inquiry* 10: 2-10.

Holmes, D. and S. Murray. 2011. "Civilizing the *Barbarian*: A Critical Analysis of Behaviour Modification Programs in Forensic Psychiatric Settings." *Journal of Nursing Management* 19: 293-301.

Holmes, D., S. Murray, A. Perron and J McCabe. 2008. "Nursing *Best Practice Guidelines*: Reflecting on the Obscene Rise of the Void." *Journal of Nursing Management,* 16: 394-403.

Holmes, D., S. Murray, A Perron and G Rail. 2006. "Deconstructing the Evidence-Based Discourse in Health Sciences: Truth, Power, and Fascism." *International Journal of Evidence-Based Health Care* 4: 180-186.

Holmes, D., P. O'Byrne and D. Gastaldo. 2006. "Raw Pleasure as *Limit Experience*: A Foucauldian Analysis of Unsafe Anal Sex between Men." *Social Theory and Health* 4: 319-333.

Holmes, D., P. O'Byrne and SJ Murray. 2010. "Faceless Sex: *Glory Holes* and Sexual Assemblages." *Nursing Philosophy* 11: 250-259.

Holmes, D., T. Rudge and A. Perron. Eds. 2012. *(Re)Thinking Violence in Health Care Settings*. Surrey: Ashgate.

Holmes, D. and D. Warner. 2005. "The Anatomy of a Forbidden Desire: Men, Penetration and Semen Exchange." *Nursing Inquiry* 12: 10-20.

Murray, SJ and D. Holmes. Eds. 2009. *Critical Interventions in the Ethics of Health Care*. Surrey: Ashgate.

Perron, A., JD Jacob and D. Holmes. Forthcoming. "The Politics of Threats in Correctional and Forensic Settings: The Specificities of Nursing Research." In *Experiencing Methodology: Narratives in Qualitative Research* edited by J. Kilty, M. Felices-Luna, and S. Fabian. Vancouver: UBC Press.

Rudge, T. and D. Holmes Eds. 2010. *Abjectly Boundless: Boundaries, Bodies and Health Work*. Surrey: Ashgate.

Selected Works

O'Byrne, P. and D. Holmes. 2012. "Indulgence, Restraint and the Engagement of Pleasure: Inciting Reflection Using Nietzsche's Ascetic Ideal." *Research and Theory for Nursing Practice* 26 (1): 1-15.

Holmes, D. and P. O'Byrne. 2012. "Resisting the Violence of Stratification: Imperialism, *War Machines* and the Evidence-based Movement." In *Evidence Based Healthcare in Context: Critical Social Science Perspectives* edited by A. Broom and J. Adams. Aldershot: Ashgate.

Rudge, T., D. Holmes and A. Perron. 2011. "The Rise of Practice Development with/in Reformed Bureaucracy: Discourse, Power, and the *Government* of Nursing." *Journal of Nursing Management* 19(7): 837-844.

McCabe, J. and D. Holmes. 2011. "Reversing Kristeva's First Instance of Abjection: The Formation of Self Reconsidered." *Nursing Inquiry* 18(1): 77-83.

Holmes, D., P. O'Byrne and D. Gastaldo. 2010. "Transgressive Pleasures: Conducting Research in the *RadSex* Domain." In *Sage Handbook on Qualitative Health Research* edited by Ivy Bourgeault. Thousand Oaks: SAGE.

Rail, G., S. J. Murray and D. Holmes. 2010. "Human Rights and Qualitative Health Inquiry: On Biofascism and the Importance of *Parrhesia*." In *Qualitative Inquiry and Human Rights* edited by N. K. Denzin & M. D. Giardina. Walnut Creek, CA: Left Coast Press.

Holmes, D. and P. O'Byrne. 2010. "Subjugated to the 'Apparatus of Capture': Self, Sex and Public Health Technologies." *Social Theory and Health* 8(3): 246-258.

St-Pierre, I. and D. Holmes. 2010. "Mimetic Desire and Professional Closure: Towards a Theory of Intra/Inter Professional Aggression." *Research and Theory for Nursing Practice* 24(2): 57-72.

10

Trevor Hussey

Emeritus Professor of Philosophy

Buckinghamshire New University, UK

1. How were you initially drawn to philosophical issues regarding nursing?

While teaching various aspects of philosophy on psychology and sociology degrees, I was offered the opportunity to teach on a new nursing degree. The enlightened programme director proposed not just the predictable ethics course but also a wider philosophical component. The experience was a strange mixture, being both sobering and exhilarating. The students were mainly experienced nurses. They were bright and keen and, between them, they brought to discussions a range and richness of experience that soon made me abandon the mundane examples I had concocted.

My task was to learn from this immersion in a new world while staying sufficiently independent to enable me to observe objectively and contribute from a philosophical perspective. I came to see that nursing discipline and practice were a uniquely complex and subtle combination of skills, knowledge, commitments, values and beliefs that provided more than enough opportunity for philosophical examination.

2. What, in your view, are the most interesting, important, or pressing problems in contemporary philosophy of nursing?

It is notoriously difficult to define philosophy and this is not the place to make an attempt. However, in clarifying my perspective I will say a little about how I see the activity. Philosophers ask, and even try to answer, the most general and fundamental questions possible. 'What do we know about the causes of strokes?' is a scientific question, but 'What is knowledge?' and 'What is a cause?' are philosophical. These kinds of question are fundamental because they underlie all our thinking. Philosophers question the assumptions underlying all disciplines; they analyse important concepts and critically evaluate people's arguments and reasons. But philosophy can also be constructive: developing very general theories, such as realism, idealism and dualism, which try to set out the ultimate nature of reality or which offer a foundation for morality, and so on. Of course, in nursing philosophy, and other kinds of applied philosophy, the questions arise more directly from issues and problems within the specific activity. There is no limit to what philosophers can investigate but concerning nursing I see four main groups of topics that are most pressing.

First, there are those issues that arise from the fact that nursing involves close, personal interactions between human beings, but interactions that take place in special circumstances. One party to the relationship is often vulnerable and in great need; they may be in pain, frightened, unconscious, physically disabled, or even dying or dead. The other party is generally a paid professional working within the constraints of an institution, with responsibilities and duties to others and sometimes encased in bureaucracy.

Out of this come issues of different but related kinds. There are moral issues about honesty, respect, the propriety of relationships, fairness, rights and decisions about priorities. There are conceptual questions concerning the way we think about personal interactions. For example, are our concepts of 'sympathy' and 'empathy' changed when used in the special relationships between nurses and patients, or do they work in the same way as in other areas of our lives? What is involved in our concept of 'choosing', 'deciding' (by both patients and nurses), 'responsibility', and 'being informed' within the special circumstances found in health care?

In applied philosophy it is often not enough to clarify concepts, we may need to bring out the implications for practice. The recent debate about spirituality illustrates this activity (Hussey 2009; Paley 2008a, 2008b; Pesut 2008a, 2008b; Newsom 2008). One area in which philosophical investigation could assist psychological and sociological work, concerns the nursing of elderly people and those with dementia. This area is of growing significance and presents problems about respect, control, choice, privacy and interactions between strangers and between different generations. For example, what do our notions of 'respect' and 'choice' mean in situations where a senile patient is being cared for in a commercial institution?

Sometimes things go wrong in nursing. There are cases of neglect, incompetence and even maltreatment and cruelty. Getting clear about empathy and sympathy, and understanding how a group culture can generate ideas about what is fair, what demands are reasonable, and what is appropriate behaviour, might help us understand how things can go so wrong.

The second group of topics I highlight are those concerned with political, economic, moral and social issues, although there are obvious relationships to the first group. Over recent decades there has been a drift towards an ever-greater influence of political and economic theories that centre on markets, consumers, commercialisation and competition.

This is well established in the USA but is growing in Europe, including the UK, and is likely to bring about profound changes in the provision of health care. Indeed, worldwide, the influence of commercial medicine and huge pharmacological companies is immense. Economists and sociologists, as well as those working in the health care professions, need to examine these developments and, for philosophers, the theories, values and concepts involved need concerted scrutiny.

This is especially true when concepts and theories are moved from their natural home, in this case commerce and manufacture, to new applications such as health care or education. I have discussed the concept of 'efficiency' (Hussey 1997) and 'students-as-customers' (Hussey and Smith 2010) in this respect but there are many others. Is a patient a customer, and does 'customer' mean the same when going into a hospital for an operation as it does when going into a shop for a pair of socks? Does the idea, assumed by much economic theory, of people as rational agents busy making free choices, make sense in health care? It has been challenged even in economics and psychology (see for example Kahneman 2011; Thaler and Sunstein 2008; Wilkinson and Pickett 2009). These issues widen out into a debate about whether our reasons are based on our desires – a theory that is presently under criticism (Parfit 2011).

Most importantly, we need a fundamental investigation of the nature of health care provision in a civilised society. Do people have a right to adequate medical and nursing care and does the ability to pay effect this right? We need to examine what our individual and collective rights, duties and responsibilities are and what constitutes social justice. Nurses are an essential part of this debate and must make their views known. My tentative contribution here is clearly from a perspective not shared by all (Hussey 2012). What should be our priorities in times of austerity? What responsibilities does the developed world have towards those in the under-developed? Of course there are also moral issues surrounding specific topics such as euthanasia, abortion, codes of ethics and so on, but they are so widely discussed that I will not stress them here.

The third group of topics concerns rationality, reason and science. When I first began to read the works of nurse researchers, theorists and philosophers I was shocked by the views of a large proportion of them. They displayed a deep distrust, almost amounting to a rejection, of science; a dismissive attitude towards reason and rationality, and a tendency to accept a range of theories and ideas that I considered profoundly dubious. There seemed to be a widespread and uncritical acceptance of highly questionable relativist theories, especially of the fashionable postmodern varieties. Distinctions between truth and falsity, and knowledge and belief appeared to be dismissed without argument. Any research that involved quantitative measurement and numbers was liable to be branded as "reductionist" or "positivist" and so worthy only of contempt. Some schools of nursing seemed to treat philosophy as the province of gurus: making the theories of a revered teacher the basis of nurse education. One example, the work of Martha E. Rogers, seemed to me to be a bizarre exercise in pseudo-science, yet it was widely lauded as an important contribution to science and to nursing (Rogers 1970).

These attitudes seemed to have infected what nurses wanted to do. It appeared that for many the only worthwhile research was qualitative, often inspired by some strange interpretation of the phenomenology of Husserl or Heidegger – and this despite admitting that much of this research could not be generalised beyond the participants involved (Paley 1997, 2000). Even worse, some theorists and nurses were embracing practices, such as therapeutic touch and some "alternative" therapies that could not be justified either by evidence or rational argument. Some theorists introduce a whole menagerie of terms, such as 'life energy', 'human energy', 'energy fields', chakras', 'cosmic unity' and 'vibrations', that are near to meaningless and certainly demand rigorous philosophical and scientific justification (Glazer 2001; Singh and Ernst 2008; Sokal 2008). I found this retreat into unreason depressing and feared that it would bring philosophy of nursing, and even nursing itself, into disrepute.

Here I must stress that I am not dismissing phenomenology, phenomenological research, qualitative research or relativism out of hand. They have contributions to make and there are certainly important philosophical debates to be had concerning them. But, for example, if small scale qualitative studies are worthwhile, then we should be able to explain why and tease out what it is about them that is useful. If we are going to embrace relativism we need to specify what sort it is, how it can be justified and what are its consequences (Baghramian 2004). All ideas, including New Age fashions, alternative therapies, mystical insights and pseudo-science etc., may be heard, but it is essential that they are subjected to rational scrutiny. Ideas have power and they are not always used for good: history shows they can cause us to do anything from waste time to kill people. What we need are reliable, rationally justifiable decision procedures to sort out the diamonds from the dross. Philosophers must ask searching questions about ideas and theories, especially the most fashionable and seductive (Hussey 2004).

Science permeates all our lives and is central to our culture. It is the best means we have yet developed for gaining knowledge and understanding of ourselves and the world. Scientific research gives us truth claims that are always tentative or probabilistic, but science also offers us objective and rationally justifiable ways of testing those truth claims and choosing between them when they conflict. These modest methods have given science amazing success and effectiveness, and given it a central role in transforming medicine and nursing for the better. This is in stark contrast to other purported methods of gaining knowledge, some of which make claims to absolute truths, yet offer no credible way of deciding which they are or of settling disputes between them.

This is not to accept science as beyond question: philosophical issues abound. We have no wholly satisfactory account of scientific method or how we can demarcate between science and non-science; there are disputes about whether science makes real progress; there are doubts about whether it gives us genuine knowledge and what "genuine knowledge" might be; questions about the nature and status of its theories or what constitutes an explanation, and so on. There are also moral issues. The standards of honesty and openness expected amongst scientists are as high, or higher, than in any other human activity, but the status of these standards is philosophically puzzling. Scientists also have to consider the moral implications of their methods, the topics of enquiry they choose, and the use made of their products. Again, there are plenty of opportunities for philosophers.

Nurses employ the fruits of science and technology every day in their work. The apparatus and instruments they use and the treatments they help to administer are generally the result of scientific research. Clearly, nurses and nurse theorists need to be critical about the changes that science and technology bring into their working lives, and they must defend the interests of their patients in the face of intrusive procedures. But this has to be done from an informed and rationally defensible stance; not from a fog of energy fields. Nurses need to conduct rigorous research that can produce conclusions that are generalisable and useful – as many already do. Nursing philosophers need to explore the implications of this approach and bring out its limitations. The idea of evidence based practice is important and seems laudable, but if there are objections to it let them be explored.

It is tempting to speculate about why nurse theorists so often display antagonism towards such things as science, logic, reason and rationality. It might be suggested that a mainly female profession is reacting against traditional domination by a largely male medical profession, which is obsessed by science, rationality and the "medical model" of human beings. Thus, in trying to establish a philosophical foundation as different from medicine as possible, nurses are calling upon the softer, romantic, intuitive dispositions of the female mind. But, as well as involving insulting caricatures of women and their minds, these sociological and psychological explanations are deeply unsatisfactory. We need a careful philosophical examination of our concepts of 'reason' and 'rationality', especially as they apply to thinking about the human relationships involved in caring. What is the relationship between rationality and empathy, sympathy and emotional responses? For example, the terms 'reason' and 'reasonable' have a range of meanings, from a reference to narrow practice of deductive logic to the broadest ideas of acting in a manner that can be justified in some way, but what are the standards of reasonableness in the situations found in nursing? What should count as a justification for what a nurse does? Do our feelings and emotions conflict with reason or are they a necessary combination? In what ways do sympathy and empathy depend on cognition and reason?

The fourth group of topics includes debates about the nature of nursing and an evaluation of various philosophical perspectives to see what relevance these may have. This would include the once fashionable activity of developing nursing "models"; inventing nursing theories, and assessing the relevance of the work of individual philosophers from Aristotle to Zeno.

Although the topics in this group are well worn they remain important, not least because they are taken up and used by nurse educators. The philosophical perspectives may be used explicitly or implicitly to colour the outlook of student nurses when they are most impressionable. As pointed out above, concerning attitudes towards science, reason and research, there may be important practical implications for the conduct of nurses, and not always for the good. I do not exclude developing novel theories specific to nursing, but it is essential that they are open to rigorous scrutiny.

My recommendation here is that we eschew treating philosophers as gurus and focus on evaluating ideas and theories critically to see their strengths and weaknesses, as well as their usefulness. My own efforts in this respect have concerned the relevance of philosophical realism, evolution and naturalism (Hussey 2000, 2002b, 2011). At a less rarefied level there are important debates to be had concerning the role of the nurse and the relationship between nurses, physicians, managers and others (for examples see Kuhse and Singer 2006, Part X).

3. What, if any, practical and/or socio-political obligations follow from studying nursing from a philosophical perspective?

Traditionally, philosophy has been seen as a "higher order" activity: one conducted at a level of abstraction and generality that separates it from the immediate practicalities of the subject it is considering. Hence there was no expectation that it would produce results that were of direct and immediate use but, it was claimed, philosophical ideas could, and often did, have a profound effect in the long term. This view has been particularly prevalent in the West and amongst analytic philosophers. I am not an historian but it seems to my amateur eyes that this picture is fairly accurate, although sometimes the delay is fairly short. Perhaps the work of political and moral philosophers has generally made an impact most quickly: Locke, Bentham, J. S. Mill, Marx, Hayek and Rawls, saw some of their ideas widely discussed or applied in their lifetimes. Even highly technical work in philosophical logic and symbolic logic, such as that of Frege, Peano, Russell and Turing, rapidly influenced the development of mathematics and modern computing.

However, over the last half century there has been a growing acceptance of "applied philosophy". This involves recognition that the skills, perspectives and procedures of philosophy can be exercised on matters of direct significance to people in their every-day life, or to the issues that arise in particular activities and professions. As Cohen and O'Hear (1984) argue in the editorial to the first issue of the *Journal of Applied Philosophy*, this is not to imply that philosophers have a monopoly of human wisdom or morality, or that philosophers are special authorities or experts; it is to conceive a philosopher as "an honest and open seeker after truth" (3). They also point out that this does not mean that the only way problems can be solved is by dispassionate, detached, neutral study: philosophers can argue from a position of commitment and concern. What is important is that any position is justified by sound and honest argument, and supported by evidence.

Things as important as nursing and health care raise vital moral, political and conceptual questions. These activities are concerned with our fundamental rights and duties to others, and they take place in a political and economic context, so that philosophers cannot avoid being involved. For example, in the U. K. the drift towards the privatisation and marketization of the health service constitutes a change of huge significance, not just to the service, but to our social culture. It will be inexcusable if philosophers do not contribute to the debate. If this is so, it is even more pressing that we engage in debates about the provision of nursing care and health care to all who need it in the developing world.

One obligation that philosophers of nursing have is to the proper education of nurses and, as the earlier comments concerning attitudes towards science, reason and research illustrate, what is taught may have a profound effect upon what nurses do in practice. Philosophy can provide help in developing critical thinking skills, an openness to ideas and a willingness to subject them to rational evaluation. To justify its place in the weighty curriculum of nurse education, philosophy has to be useful.

4. In what ways does your work seek to contribute to philosophy of nursing?

My approach to philosophy is in what can loosely be called the analytic tradition. I try to offer carefully argued and rationally justified contributions of three main kinds. First, an examination of important or fashionable concepts, in which I try to explicate their meaning and draw out their implications for nursing (Hussey1997, 1998, 2002a, 2009). Second, an examination of philosophical or scientific theories, with the aim of investigating their usefulness in understanding the practice of nursing as it is, or could be (Hussey 2000, 2002b, 2011). Third, papers in which I argue for a specific thesis, such as a particular view of social justice as it applies to health care (Hussey 2012), or to urge a more rational approach in philosophising about nursing (Hussey 2004).

5. Where do you see the field of philosophy of nursing to be headed, including prospects for progress regarding the issues you take to be most important?

Prediction is very difficult: a wish list is easier. So much depends on what happens in the health services around the world. If the trend towards commercialisation continues we may find that educated nurses become too expensive and are replaced, in part at least, by trained care assistants. This may then affect the schools of nursing and the curricula they offer, and philosophy may be seen as a luxury. If this gloomy prospect does not materialise and the nursing profession continues to grow in confidence, philosophy may prosper. In either case the need to think deeply about the nature of nursing, its responsibilities and its principles will be just as vital.

There will always be a need to critically examine fashionable concepts. One example is 'holism'. At the present, no philosopher would dare to produce a theory of nursing which did not claim to be holistic: it is mandatory. But what does 'holistic' mean and what are its implications? The literature is vast, but a casual scan shows that there is a wide range of meanings. At one extreme 'holistic' seems to suggest a tick list that covers everything a nurse should take into account when caring for a patient – from their spiritual wellbeing to clipping their toenails, via their ethnic origin and their relationship with their father. At the other end holism disappears into a mystical fog of universal interconnectedness, cosmic consciousness, and vague claims about wholes being greater than the sum of their parts. Quite how a nurse is supposed to do a specific task for a particular patient with a singular problem at a certain time, is unclear.

If nurses and philosophers of nursing want their profession to progress and their patients to be better served, what matters in the long term is that they have the intellectual curiosity to take up the broad questions concerning their activities, and pursue them with open minds and critical rigour.

References

Baghramian, M. 2004. *Relativism*. London: Routledge.

Cohen, B. and A. O'Hear. 1984 "Editorial: A Note on Policy." *Journal of Applied Philosophy* 1: 3-4.

Glazer, S. 2001. "Therapeutic Touch and Postmodernism." *Nursing Philosophy* 2: 196-212.

Hussey, T. B.

1997."Efficiency and Health." *Nursing Ethics* 4: 181-190. Reprinted in (1998) *Monash Bioethics Review* 17: 12-21.

1998."Change and Nursing." In *Philosophical Issues in Nursing* edited by S. Edwards, 47-64. Basingstoke: MacMillan.

2000."Realism and Nursing." *Nursing Philosophy* 1: 98-108.

2002a."Thinking about Change." *Nursing Philosophy* 3: 104-113.

2002b."Evolution and Nursing." *Nursing Philosophy* 3: 240-251.

2004."Intellectual Seductions." *Nursing Philosophy* 5: 104-111.

2009."Nursing and Spirituality." *Nursing Philosophy* 10: 71-80.

2011."Naturalistic Nursing." *Nursing Philosophy* 12: 45-52.

2012."Just Caring." *Nursing Philosophy* 13: 6-14. First given as keynote address to IPONS conference '*Philosophizing Social Justice in Nursing*.' Vancouver 2010

Hussey, T. B. and P. Smith. 2010. *The Trouble with Higher Education*. London: Routledge.

Kahneman, D. 2011. *Thinking, Fast and Slow*. London: Allen Lane.

Kuhse, H. and P. Singer. eds. 2006. *Bioethics: An Anthology*. Oxford: Blackwell.

Newsom, R. W. 2008. "Comments on 'Spirituality and Nursing: A Reductionist Approach" by John Paley. *Nursing Philosophy* 9: 214-217.

Paley, J. 1997. "Husserl, Phenomenology and Nursing." *Journal of Advanced Nursing* 27: 187-193.

Paley, J. 2000. "Heidegger and the Ethics of Care." *Nursing Philosophy* 1: 64-75.

Paley, J. 2008a. "Spirituality and Nursing: A Reductionist Approach." *Nursing Philosophy* 9: 3-18.

Paley, J. 2008b. "Spirituality and Nursing: A Reply to Barbara Pesut." *Nursing Philosophy* 9: 138-140.

Parfit, D. 2011. *On What Matters*. Oxford: Oxford University Press. Volume 1.

Pesut, B. 2008a. "A Conversation on Diverse Perspectives of Spirituality in Nursing Literature." *Nursing Philosophy* 9: 98-109.

Pesut, B. 2008b. "A Reply to 'Spirituality and Nursing: A Reductionist Approach' by John Paley." *Nursing Philosophy* 9: 131-137.

Rogers, M. E. 1970. *An Introduction to the Theoretical Basis of Nursing*. Philadelphia: F. A. Davis Company.

Singh, S. and E. Ernst. 2008. *Trick or Treatment: Alternative Medicine on Trial*. London: Bantam Press.

Sokal, A. 2008. *Beyond the Hoax: Science, Philosophy and Culture*. Oxford: Oxford University Press.

Thaler, R. H. and C. R. Sunstein. 2008. *Nudge: Improving Decisions about Health, Wealth and Happiness*. Yale New Haven: University Press.

Wilkinson, R. and K. Prickett. 2009. *The Spirit Level: Why More Equal Societies Almost Always Do Better*. London: Allen Lane.

11

June F. Kikuchi

Professor Emeritus, Faculty of Nursing

University of Alberta, Edmonton, Canada

---

1. How were you initially drawn to philosophical issues regarding nursing?

Pieper's (1952) characterization of philosophizing best describes the impetus behind my study of philosophical issues regarding nursing. According to Pieper, wonder about the world – the beginning of philosophical thought – requires a backing away from the workaday world in which things are narrowly viewed in pragmatic terms with the aim of finding solutions to problems. However, we cannot simply tell ourselves to back away. A shock is required that awakens us from a complacency of which we are unaware to a realization that we do not understand as much as we think we do. Socrates' questions delivered such a shock. He "compared himself ... to an electric fish that gives a paralyzing shock to anyone who touches it" (98). The electric fish that shocked me out of my complacency was Dr. Helen Simmons, a University of Alberta Faculty of Nursing colleague with graduate preparation in philosophy.

After practicing as a general duty nurse, a head nurse, and a clinical nurse specialist, I took on a nursing academic role at the University of Alberta on completion of my PhD studies in Nursing in 1979. A couple of years later, the Faculty of Nursing decided to design a graduate program in community health nursing. During the first planning meeting, Helen Simmons asked the question, "What is nursing?" Her rationale for the question was: it is only by defining nursing that we will know we are designing a *nursing* program and not, for example, a *medical* program. That made sense; so, discussion about course offerings ceased, as we attempted to answer the question. When time began to run short with no adequate answer in sight, we returned to the task at hand: to identify course offerings. However, at the next meeting, Helen again asked, "What is nursing?" Once again, we attempted to answer the question; once again, we failed.

Immersed in the workaday world, I saw Helen as an obstacle to our planning: she was continually moving us away from our task. When I learned that Helen was not a nurse, I thought: "All nurses know what nursing is and were she a nurse, she would not need to ask." But then, I wondered, "Why could we not explain it to her adequately. Why were we stumbling around?" I concluded that lack of thinking time was the problem and that, if I gave her question sufficient time, I could answer it. After all, I had a PhD in *Nursing* degree and had spent years as a practicing nurse. My goal was to give Helen an answer that would satisfy her, so that we could get on with our planning. When, however, try as I might, I could not define nursing adequately, I was startled and shaken. As Pieper (1952) would say, the dome of my workaday world was pierced. I was shaken in the way my undergraduate philosophy course had shaken me but much more so. I had been fooling myself. I had taken for granted that I understood what I did not understand.

I revealed my discovery to Helen and was pleasantly surprised with her invitation to explore the nature of nursing with her. She explained that definitions are a philosophical matter, so we would need to philosophize. Up to then, I had not thought of philosophy as a mode of inquiry. Like many others, I was under the impression that there was only one mode of inquiry – the scientific. Scared but excited about deepening my understanding of nursing and philosophizing, I accepted Helen's invitation. Doing so proved to be life changing.

Once into our research in the early 1980s, I began to realize that serious philosophical study requires lengthy concentrated periods of time with little "pay off' in terms of academic advancement since opportunities to obtain research grants and to publish in nursing philosophy were then almost nonexistent. But there was another kind of pay off that was more alluring: eradicating my ignorance about the nature of nursing. As I ploughed ahead, I discovered that an understanding of the nature of nursing entailed an understanding of such basic matters as the nature of reality, the human mind, and free will. Before long, I was immersed in metaphysics, epistemology, ethics, and politics and in the epistemological issues emerging within the nursing discipline. Then, serendipitously, Dr. Shirley Stinson, Associate Dean of the Graduate Nursing Program, asked Helen and me to develop and teach a philosophy of nursing course for the master's of nursing (MN) program. Needless to say, we were ecstatic but scared.

Teaching MN students to philosophize was a challenge for them and us. Accustomed to thinking concretely, the students found it difficult to think abstractly and to understand the necessity of philosophizing. But when they discovered how little they understood the nature of nursing and of human beings and the basis of their ethical decisions, they were hooked and eager to learn. Before long, Helen and I began to discuss the importance of bringing philosophical thinking out of its embryonic state within the nurse's world in order to develop the philosophical knowledge underlying the practice of nursing. One thing led to another, and in 1989, Helen and I co-founded the Institute for Philosophical Nursing Research at the University of Alberta. With its establishment, Helen and I implemented various ways to bring nurses together to philosophize about their world: conferences devoted to philosophical discussion about issues facing the nursing discipline and profession, publications, open lectures, workshops, and so forth.

2. What, in your view, are the most interesting, important, or pressing problems in contemporary philosophy of nursing?

In my view, a chief pressing problem in contemporary philosophy of nursing is the lack of clear, coherent and true (in the sense of true beyond a reasonable doubt) philosophical conceptions of *nursing* and *philosophy*, and consequently of *philosophy of nursing* to guide its development. What is philosophy of nursing? What distinguishes it from other endeavors? How do we determine if our work or that of others falls within the domain of philosophy of nursing? Problematically, such definitional questions have fallen by the way side in contemporary philosophy of nursing. The struggles of the latter part of the 20th century to define nursing and nursing knowledge have for the most part not been carried into the 21st century.

When I ask nurses working in contemporary philosophy of nursing to define nursing, philosophy, and philosophy of nursing, I am usually met with a bemused look. I am quickly reminded of how past attempts to define nursing and nursing knowledge proved to be futile and that seeking definitions is a thing of the past. However, when they are then asked for the basis of their claim that they are working in philosophy of nursing, they usually begin to tackle the question as though it were an easy one to address. On discovering that is not the case, they become defensive. They get angry and say that it is a waste of time to go down the path of seeking definitions. The conversation ends with more heat than light having been generated – yet another pressing problem.

I suspect that, these days, those who attempt to converse about the nature of philosophy of nursing have been or will be met with resistance and an abrupt end to the conversation. In contemporary philosophy of nursing, it is becoming difficult and at times impossible to have amicable discussions even about non-definitional matters. Typically, when adherents of a particular perspective (A) question or criticize a contrary perspective (B), adherents of perspective B immediately think that their perspective is under attack and develop a defensive posture. The possibility that the question or criticism flows from a genuine desire to understand or help improve the perspective is not entertained. Derogatory comments start to fly. For example, each party questions the integrity and intelligence of the other. Eventually, the fighting dies down only to flare up later somewhere else.

What is behind this non-scholarly behavior? In discussing such behavior, Johnstone (2012) says that what we have is "(a) problem of ideological judgments masquerading as scientific or discipline judgments", possibly resulting from "the decline of truth as a social value generally" (111). I would go one step further and say that relativism is the engine driving the decline of scholarly behavior in contemporary philosophy of nursing. With epistemology rather than metaphysics being foundational in relativism, knowing reality as it exists independent of the individual mind (i. e., objective reality) becomes moot: reality is a reflection of the mind (i. e., a construction of the mind). It is what we subjectively think it is. Given this conception of reality, the following related notions are repeatedly heard: perspective is the "creator" of reality; knowledge of objective reality is unattainable; and, there is no objective measure of the truth of a perspective. The upshot is that mere opinion is accepted as knowledge and the most that one can hope for is an exchange of ideas. In the absence of a common objective arbitrator of truth (i. e., objective reality), discussion of ideas to arrive at the truth of a matter is pointless. So then, an important question arises: "What is the purpose of philosophy of nursing nowadays?"

A look at recent philosophy of nursing literature might lead one to think that philosophy of nursing has transformed itself into a socio-political science and its purpose is to eliminate inequities in health care through socio-political research and political activism. With nurse philosophers behaving like socio-political scientists and political activists, I cannot help but wonder: "Why are they studying socio-political science questions and moving into political activism? What is philosophical about their activities?" In raising these questions, I am not saying that the work they are doing is unimportant. I credit them with increasing our awareness of how inequities are being created by the current socio-economic-political climate and policies and of the need to engage in political activism to alter this state of affairs. But, in so doing, they are not philosophical. Like political activists, they start from a stance that they unquestioningly adopt as correct. They do not ask philosophical questions about the nature of justice, liberty, and equality. They merely tell us what they are and use rhetorical devices to have us join their bandwagon.

So, then, why are nurse philosophers studying socio-political science questions and moving into political activism? I suspect that the answer is multifaceted. The move into socio-political science may be the influential effect of nurses who have studied in the social and political sciences. It may also have to do with the ongoing confusion between science and philosophy. The move into political activism may be related to the replacement of the pursuit of truth and knowledge with the pursuit of political power.

3. What, if any, practical and/or socio-political obligations follow from studying nursing from a philosophical perspective?

The philosophical ground and research question shape what, if any, obligations follow from studying nursing philosophically. The philosophical ground, the suppositions underlying the study (i. e., its presuppositions), sets down the basis for the existence and nature of obligations and the conditions under which they exist. The research question then shapes the obligations related to the study. Let me demonstrate my point by answering the posed question from the philosophical perspective of moderate realism.

Given the space limitations, only the basic underpinnings and tenets of moderate realism most relevant to the discussion are identified. In moderate realism, it is held that reality exists independent of the individual mind. Human beings attain knowledge of reality through the cooperative efforts of their immaterial mind and material senses. Given their common human nature, all human beings possess the capacity to choose freely and all have natural needs such as the need for food, sleep, friends, knowledge, peace, dignity, protection from harm, and so forth. By virtue of their natural needs, all human beings possess natural rights to the necessary goods to meet their needs. However, not all human beings possess legal rights as they are rights conferred by society. Rights in turn place certain obligations on human beings. In acting for the common good, human beings benefit their own good (Adler 1971).

According to Adler and Van Doren (1972), in moderate realism, philosophy is considered to be a mode of inquiry that, unlike science but like mathematics, is conducted from the armchair and does not collect data for research purposes. As such, in philosophy, there are no obligations related to the collection of data as there are in science (e. g., the obligation to secure informed consent). Philosophy and other modes of inquiry share a common object: to seek knowledge in the sense of probable truth (i. e., true beyond a reasonable doubt). The specific object of philosophy is to seek *philosophical* knowledge. Being bound by that object, philosophers have a right to the necessary goods (e. g., freedom of speech) to attain it and an obligation not to obstruct others' access to those goods. If a conclusion is based on compelling evidence and reason, philosophers as well as others are obligated to agree with or accept the conclusion; if not, not. Agreement with a theoretical conclusion carries no obligation to act; but, agreement with a practical conclusion carries an obligation to act accordingly. In the case of philosophers, that obligation is circumscribed by their object to seek philosophical knowledge. In other words, practical or socio-political activities unrelated to the seeking of philosophical knowledge are not their responsibility. On the other hand, they may be obligated to engage in such activities in other capacities (e. g., as practitioner, citizen, politician). An example may help clarify this point.

Suppose a philosophical study is undertaken to determine if the following practice should be continued: the provision of nursing care by non-licensed non-regulated workers with little if any training. And, suppose it is determined beyond a reasonable doubt that it should not be continued because it is unjust. Workers in health care and other relevant persons would be obligated to help discontinue the practice, prudently considering such variables as their roles, their agencies' policies, and the socio-political-economic climate. However, the researcher of the study would not incur that obligation in her/his capacity *as researcher*, although s/he may do so in another capacity (e. g., as politician).

Finally, philosophical research is necessary to determine whether an entity is a legitimate right or obligation. Unfortunately, in contemporary philosophy of nursing, this is not the prevalent view. Rather, under the influence of relativism, it is held that if you think X is a nursing obligation and I think it is not, we are merely expressing our different views of reality, desires, and tastes. Under these circumstances, there is no room for philosophical research. If you think that X is a nursing obligation, you can make it so by getting people on board with your view. The more people get on board the better, because "might makes right." Lipscomb (2011) draws our attention to the appearance of this kind of thinking in nursing philosophy with regard to social justice, as does Ballou (2000) in relation to sociopolitical activism and professional codes of ethics. They note that nurses are being made responsible for social justice according to a favored philosophical theory of domination, freedom, and so forth.

4. In what ways does your work seek to contribute to philosophy of nursing?

In the early 1980s when I began my work in nursing philosophy, my aim was very personal: to deepen my understanding of philosophy. On realizing the potential that lies in philosophizing about nursing to make a difference that makes a difference, my aim expanded to helping nurses understand the place of philosophy in their world. At that time, philosophical inquiry was in an embryonic stage in nursing. There was no formal venue where nurses could discuss philosophical nursing matters with one another. Philosophical study in nursing was taking place sporadically and was rarely included in graduate nursing programs. It was little known, recognized, and reported. To wit, when Helen Simmons and I looked for philosophical nursing literature to include in a graduate philosophy of nursing course on the nature of nursing, the pickings were slim. Literature on nursing theories was plentiful; but little of it was philosophical in nature. Although there was an eagerness to develop theories of the nature of nursing, there was little understanding of the need to think philosophically to do so.

Helen and I began to think about what would be required to advance the philosophical study of nursing beyond its embryonic stage. Before long, we were envisioning "a home' dedicated to the philosophical study of nursing – a place that would bring together nurses who wanted to discuss and study philosophical nursing questions and learn about the place of philosophy in the nurse's world. With help from members of the University of Alberta Faculty of Nursing, the Winspear Foundation, and other supporters, that home became a reality. In 1988, with a shoestring budget and permission from the University of Alberta, Helen and I co-founded the Institute for Philosophical Nursing Research (IPNR). The following year, under our directorship, the Institute hosted its first biennial "Philosophy in the Nurse's World" conference in Banff, Alberta. The conference was designed to promote philosophical discussion of the conference topic "Philosophic Inquiry in Nursing" through ample time for discussion, limiting participation to 40 nurses, and encouraging disciplined discussion. The conference site and format worked their magic – participants were soon philosophizing in the sessions and afterwards in the mountains. Word quickly spread about the conference and requests to attend future conferences were received. Papers presented at the conference were published by Sage. That first conference set the standard for the subsequent conferences we organized until our retirement as IPNR directors in 1997.

Other means of engaging nurses in the philosophical study of nursing were developed by the IPNR under Helen's and my directorship (e. g., workshops, working groups). Their impact however was more local in contrast to the global impact of the IPNR's conferences and subsequent publications. The latter helped bring philosophy of nursing out of the shadows. They came on the scene at the right time. Nurse educators, researchers, scholars, and clinicians were ready for them. They were struggling with philosophical questions about nursing and eager to discuss them with one another and to read philosophical nursing literature that addressed their questions. Most importantly, they had begun to realize that the scientific mode of inquiry was not the only mode of inquiry available to them.

Once philosophical inquiry caught hold in nursing, it seemed to explode. Today, philosophy of nursing conferences are being held regularly and philosophy of nursing literature is being published in several nursing journals and regularly in the journal, *Nursing Philosophy*. There is a philosophy of nursing community enabled by the International Philosophy of Nursing Society. Most, if not all, doctoral nursing programs include study of nursing theories, ethical theories, and the philosophical underpinnings of various approaches to knowledge development.

Helen and I set out to do what we could to help advance the philosophical study of nursing. Looking back at our work, it would seem that our main contribution to that effort was to ignite nurses' interest in thinking philosophically about nursing problems and to provide various ways of pursuing that interest. In all our endeavors in matters of truth, we valued diversity of thought but as a means to attaining unity of thought based on reason and evidence. This value is reflected in our publications. I continue to be guided by this value, as I am convinced that it is the aim of inquiry. Seeking only diversity of thought makes a farce of inquiry; yet, it is viewed otherwise by many in contemporary philosophy of nursing who state that unity of thought will stifle inquiry and freedom of thought. In so doing, they fail to consider the conditions under which that is likely to happen and the conditions under which unity of thought will likely do the opposite.

5. Where do you see the field of philosophy of nursing to be headed, including the prospects for progress regarding the issues you take to be most important?

My response to this question is foreshadowed in my earlier essays. I predict that, if the field of philosophy of nursing continues on its present course, it will meet the same fate as the nursing theories of the last century. It will become impotent and irrelevant and be put out to pasture. Perhaps, nurse philosophers will come to realize in time that the babble of mere opinion and the use of "might" in place of reason to settle arguable matters are doing more harm than good. But I do not hold much hope of that happening, given relativism's stronghold on nurse philosophers. So, I have wondered if my hope is better placed on practicing nurses who have their feet more firmly planted in reality. After all, they, not nurse scholars, put the nursing theories out to pasture.

Amidst the protests of nurse theorists, practicing nurses rightfully complained that the nursing theories were esoteric. They saw the danger of each nurse practicing according to her/his preferred nursing theory – a real danger that nurse theorists seemed to ignore and not address. The chances, however, that practicing nurses will rise up and take nurse philosophers to task are very slim. Whereas nursing theories were hoisted on practicing nurses, philosophy of nursing has not been.

Also, most practicing nurses are unaware of its existence. Further, since exposure nowadays to philosophy of nursing at the graduate level in nursing is mainly relativistic in nature, graduates return to practice with feet less or no longer firmly planted in reality. So, where does all of this leave us?

As nurse philosophers, we can choose to carry on as we are or we can wake up and start to make the changes that are necessary to enable philosophy of nursing to regain its potency. Of course, that is easier said than done. It will require restoration of a love of wisdom and disciplined philosophical discussion of the kind found in the dialogues of Socrates and Plato wherein a sequence of related questions and answers about a philosophical matter are examined to separate the wheat from the chaff. The dialogue proceeds in a spiral fashion; with each turn of the spiral, more chaff is removed and insights are gained. To be successful, all parties to the dialogue would have to agree to conduct themselves in ways that promote enjoyable fruitful discussion (e. g., be polite, give others their due, stay on course, avoid tangents). Most importantly, participants would have to be open-minded and prepared to go where the dialogue logically moves. Along this vein, disclosure of philosophical leanings would make for a more open dialogue (Adler and Van Doren 1972).

If the philosophical matter is definitional in nature, then a way to proceed is to start with a very general definition of the matter – one to which all can agree. Then, ask a question about it and about the answer. Other questions will arise as will other answers, revealing the issues that need to be resolved to arrive at a more specific definition. Adler (1958) used this method to identify issues that must be resolved to attain an adequate definition of such philosophical entities as justice. Needless to say, the process of identifying and then resolving the identified issues requires considerable time, patience, and teamwork. I daresay that it may be difficult to find nurse philosophers who would be willing to take on such research, given that long term research is not highly rewarded in academia. Given this situation, definitions such as that of the nature of nursing will likely continue to elude us; and, we will continue to say that if we do not define nursing, it will be defined by external vested interest groups.

Just a couple of decades ago, the chief obstacle to the advancement of the nursing profession was the medical profession. Today, the nursing profession is confronted by other professional and nonprofessional groups, hospital boards, and governmental bodies with power to change the course of the profession, and are doing so at an alarming pace. With registered (licensed) practical nurses now allowed to use the designation of nurse and to perform a great number of activities formerly carried out only by registered nurses (RNs) and with personal support workers providing "basic nursing care," RNs are wondering what will become of them, nursing care, and the nursing profession.

In conclusion, the need for a clear, coherent, and true philosophical conception of nursing is becoming more and more pressing. Meanwhile, the field of philosophy of nursing whose job it is to meet that need is doing little to meet it. Further, things will not change until nurse philosophers realize the necessity of such a conception and of disciplined philosophical dialogue to attain it.

References

Adler, Mortimer J. 1958. *The Idea of Freedom. A Dialectical Examination of the Conceptions of Freedom*. Garden City, New York: Doubleday.

Adler, Mortimer J. 1971. *The Common Sense of Politics*. New York: Holt, Rinehart and Winston.

Adler, Mortimer J., and Charles Van Doren. 1972. *How to Read a Book*. New York: Simon and Schuster.

Ballou, Kathryn A. 2000. "A Historical-Philosophical Analysis of the Professional Nurse Obligation to Participate in Sociopolitical Activities." *Policy, Politics & Nursing Practice* 1: 172-184.

Johnstone, Megan-Jane. 2012. "Academic Freedom and the Obligation to Ensure Morally Responsible Scholarship in Nursing." *Nursing Inquiry* 19(2): 107-115.

Lipscomb, Martin. 2011. "Challenging the Coherence of Social Justice as a Shared Nursing Value." *Nursing Philosophy* 12(1): 4-11.

Pieper, Josef. 1952. *Leisure: The Basis of Culture*, Translated by Alexander Dru. New York: Pantheon Books.

Selected Works

Kikuchi, June F.

1992."Nursing Questions that Science Cannot Answer." In *Philosophic Inquiry in Nursing*, edited by June F. Kikuchi and Helen Simmons, 26-37. Newbury Park, CA: Sage.

1996."Multicultural Ethics in Nursing Education: A Potential Threat to Responsible Practice." *Journal of Professional Nursing* 12(3): 159-165.

1997."Clarifying the Nature of Conceptualizations about Nursing." *Canadian Journal of Nursing Research* 29(1): 97-110.

2003."Nursing Knowledge and the Problem of Worldviews." *Research and Theory for Nursing Practice* 17(1): 7-17.

2004."Towards a Philosophic Theory of Nursing." *Nursing Philosophy* 5: 79-83.

2011."Thinking Philosophically in Nursing." In *Canadian Nursing: Issues and Perspectives*, edited by Janet C. Ross-Kerr and Marilynn. J. Wood, 5th ed., 105-117. Toronto: Elsevier Mosby.

Kikuchi, June F., and Helen Simmons. 1996. "The Whole Truth and Progress in Nursing Knowledge Development." In *Truth in Nursing Inquiry*, edited by June F. Kikuchi, Helen Simmons, and Donna Romyn, 5-18. Thousand Oaks, CA: Sage.

12

Timothy Kirk

Assistant Professor, Philosophy

City University of New York—York College, New York, USA

---

1. How were you initially drawn to philosophical issues regarding nursing?

I fell into the philosophy of nursing quite by accident. When I began graduate school in philosophy in 1994, I chose a program in Continental philosophy. I was very interested in the construction and deconstruction of meaning and value in human experience. It seemed to me that the 20th Century French and German philosophers were especially adept at opening up questions in this area in a way that was attentive to capturing the richness and complexity of human experience, a richness and complexity I did not find compellingly captured by the linguistic turn taken in 20th Century Anglo-American philosophy.

I was part of the inaugural class of PhD students at Villanova University, where I received a very strong education in existentialism and phenomenology. After my first few years of coursework, however, I found myself searching for a more concrete way to connect my education and interests in philosophy to the everyday lives of the people around me. Although focused on the nature and meaning of experience, I found that my work in existentialism and phenomenology was highly theoretical and was lacking a rootedness in experience—that is, its connection to the everyday lived experience of people I knew and encountered (including myself) appeared tenuous at best. Coincidentally, at the same time I was searching for this rootedness, one of the faculty members in my department suffered an injury. As a result, I found myself thrust into teaching his healthcare ethics class to 30 undergraduate nursing students while he recuperated for several weeks.

That few weeks changed my life in two significant ways. First, I discovered areas of scholarship in healthcare ethics and the philosophy of medicine that seemed the perfect blend of my philosophical interests with powerful human experiences of health and illness. Second, I discovered the joy of teaching and working with nurses, who had a sense of purpose and focus as students and colleagues that was different from what I had experienced working with other students and colleagues in the liberal arts.

At the time, most US scholars in healthcare ethics were focused almost exclusively on acute care medicine. What I saw in nursing, however, was a model of care—sometimes called the "biopsychosocial" model—that seemed to have obvious resonances with phenomenology and my related areas of interest philosophically. That led to a dissertation on intimacy in nurse-patient relationships. In researching and writing that project, I discovered the work of nursing theorists like Hildegard Peplau and Imogene King, both of whom had rigorous and compelling conceptual models—Peplau of therapeutic relationships (1952) and King of goal attainment (1981)—that I thought blended philosophical enquiry with implications for practice in a manner I encountered all too rarely in work written by professionally trained philosophers.

2. What, in your view, are the most interesting, important, or pressing problems in contemporary philosophy of nursing?

My own approach to the philosophy of nursing is very attentive to the practice of nursing. Indeed, one of the aspects of the field I find most appealing is that there is an almost even mix of nurses and philosophers active in its scholarship. When at its most useful (at least for me), the philosophy of nursing is addressing the fundamental assumptions, conceptual models, and conditions of practice that touch directly on the daily experience of nurses, patients and families. A danger of being a philosopher in this field is that one will choose to ask questions, or—what's worse—propose ways of addressing or answering questions, that do not reflect or contribute to nursing as it is practiced clinically. So, I tend to think of the importance of "problems" in the contemporary philosophy of nursing as being guided by what those who give and receive nursing care perceive as in urgent need of exploration and clarification.

I live and work in the U. S., where we have a healthcare system that is in crisis. We have exorbitant healthcare costs, spending far more per patient than any other country on the planet. At the same time, we have mediocre outcomes. While our median life expectancy continues to rise, the health and wellness of our population is in decline. The U. S. has a hospital-based healthcare system that is very good at rescuing patients who have acute health crises. We lack, however, the community-based infrastructure that can, through prevention and continuity of care, stem the development of chronic illness across the lifespan and address the burden of decompressed morbidity in the last decade of life.

I highlight this crisis for two reasons. First, it is increasingly contributing to significant distress amongst nurses who practice in the system. We teach nurses to engage with their patients in a way that assesses and identifies multiple contributing causes of health and illness. The four-year nursing curriculum in the United States emphasizes aspects of nursing like building relationships, teaching patients about their health and their bodies, and the importance of integrating elements such as diet, exercise, social engagement, and spiritual fulfillment into promoting wellness. Then, upon graduation these nurses enter a practice environment wherein—due to high caseloads, increasing patient acuity, decreasing length of stay, and the continued dominance of a biomedical model of illness—they spend the overwhelming majority of their time administering medications to patients and documenting that medication administration. There is, in other words, a jarring dissonance between what nursing scholars and theorists say nursing "is" (and, in turn, what nursing students are taught constitute the defining elements of nursing as a discipline) and what the conditions in which nursing is practiced allow nurses to do. To the extent that this dissonance is producing distress and is the product of a disconnect between philosophy and practice, it is an urgent problem. And, I would argue, it is not only a problem of healthcare policy or workforce development, though its effects on both are negative. It is also a problem of identifying and reducing barriers to nurses—individually and collectively—living rich, integrated lives marked by philosophical coherence and a clarity of vision that supports the pursuit of meaning and truth.

The second reason I highlight the crisis in the U. S. healthcare system is that I believe that the philosophy of nursing—by which I mean the conceptual foundations and norms of practice that explore, guide, and inform what it means to be a nurse and how nursing is able to do what it does, as well as the aspirational questions that ask what nursing could be—is uniquely capable of addressing the U. S. healthcare crisis. The conceptual re-framing and shifting of policy foci required to better address chronic health conditions, build relationships with individuals, families, and communities that promote health and wellness rather than exclusively react to acute injury and illness, and engage the multifactorial influences that sustain health and predict risk of illness—this reframing and shift in policy will need to be grounded in a paradigm that is coherent, theoretically sound, and has robust explanatory power. The philosophy of nursing—in which I include the work of nurse theorists and philosophers of nursing—has been building just such a paradigm over the past century. Continuing to develop a well formulated, clearly articulated, and theoretically robust philosophical framework to capture, explain, analyze, refine, influence, and argue for some of the core commitments and practice models at the heart of nursing will be crucial in moving toward redesigning a health care system that is economically sustainable and engages communities to produce better health outcomes for a greater percentage of the population than currently have access to care in the U. S.

Some might argue that what I am advocating/describing above is not "philosophy of nursing" but "nursing philosophy," the former being the study of the assumptions, methods, and purposes that help define what nursing "is" and the latter being what nursing as a discipline puts forth as its "approach" or "point of view" to larger questions of health and illness. In other words, the former could be seen as proper philosophical enquiry from an academic point of view while the latter might be seen as a kind of ideology or system of beliefs about how one should approach the enterprise of promoting health or addressing illness in society at large.

I would argue, however, that while this is a legitimate distinction between foci, both are properly philosophical. If one takes Sellman's (2000, 2011) application of Alasdair MacIntyre's work to nursing seriously (and I do), then there is great utility in considering nursing to be a "moral practice." In such a practice, the epistemological, ontological, ethical, and hermeneutic functions of nursing are embedded in a dynamic teleological dance between elements internal to nursing itself and the other social practices with which nursing finds itself always already embedded. As such, one cannot truly address philosophical questions about nursing if one does not take into account the larger sociopolitical context in which nurses are educated, employed, and collectively constitute a part of larger social systems devoted to addressing civic issues related to health and illness.

Insofar as this is the case, I think an important task in contemporary philosophy of nursing is to integrate the foci and methods of social and political philosophy with some of the historical strengths—epistemology, phenomenology, philosophy of science, ethics—in the field. Thinking about how to blend some of the philosophical work supporting and explaining the core commitments of nursing as a discipline—therapeutic relationality, the biopsychosocial model of health and illness, nurses as teachers promoting self-understanding and self-efficacy, the conceptual building blocks of nursing assessment and nursing care-plans—with (a) questions about the best way for communities and civic governments to promote and protect the health of their people, (b) the role of health in balance with other civic goods, and (c) the relationships between human health and economic and environmental practices (relationships increasingly supported by research in public health)—this blending will, I think, open opportunities for philosophical discovery and policy innovation that are currently dominated (when pursued at all) by medicine, which has done a much better job of securing for itself a seat at the sociopolitical table.

Given the worldwide demographic shift toward a population which will have a much higher median age and, along with that age, a greater morbidity burden in chronic illness, the time is right for nursing to assert itself as an alternative to (or, perhaps, a significant supplement to) medicine-based healthcare systems. Philosophers of nursing (or, if you prefer, nursing philosophers) would do well to devote more effort to developing the conceptual frameworks and philosophical structures that could enable the natural synergies between the fundamental commitments of nursing as a discipline and the health-related needs of populations to be uncovered and articulated in the public sphere.

3. What, if any, practical and/or socio-political obligations follow from studying nursing from a philosophical perspective?

I have addressed this rather comprehensively in the previous section, as I believe the two questions are closely related.

4. In what ways does your work seek to contribute to philosophy of nursing?

My major interests in the early 2000s focused on the interpersonal dynamics in nursing that opened up possibilities for therapeutic interactions with patients. I attempted to explore those dynamics through the lens of meaning—the meaning of illness and health for patients, the meaning of the work for nurses, and the shared senses of meaning that I suspected were required for a therapeutic interpersonal connection to open up possibilities for healing. In putting forth a conceptual model of clinical intimacy (Kirk 2007), I offered one way to explain and capture how and why some nursing relationships with patients were able to engage patients in healing and others were not.

I would love someday to operationalize and test that model. Because my approach to philosophy involves building conceptual models in an attempt to capture the lived experience of nurses and patients, and because doing so holds such promise (I think) for better understanding how the experience of both contributes to larger questions of health and illness, in 2007 I began to pursue a second master's degree. This time I am studying public health in the department of sociomedical sciences of the Mailman School of Public Health at Columbia University. I undertook this course of study to get formal training in quantitative and qualitative research methods, thinking such training would give me insight into different ways of operationalizing and assessing conceptual models like my model of clinical intimacy. At the moment [2013], I am half way through my studies and have already learned a great deal that I suspect will contribute to future work in this area.

Beginning around 2008 I began to focus more narrowly on hospice nursing. Although imported from the UK in the 1960s and 1970s, hospice has evolved differently in the U. S. than in many other countries. One way that development peaked my interest is the prominent role played by nurses in home hospice care. Although part of an interdisciplinary team that includes social workers, spiritual care providers, physicians, bereavement counselors, and others, home hospice care in the U. S. developed such that the care of patients and families became very nurse-driven. As such, it seemed to me a natural laboratory in which to explore how the influence of nursing (and its associated philosophical paradigm) impacted the experience for all involved—patients, families, and clinicians alike. And, while hospice itself originated as something of a political movement that had roots in clear, robust philosophical writings by its founder, Cicely Saunders, the synergy between the hospice philosophy and some of the core commitments of nursing noted above is powerful (Saunders was a trained nurse and physician).

Much of the work I do now seeks to explore resonances and dissonances between the philosophies of nursing and hospice care, and between the common elements of those philosophies and the sociopolitical context in which hospice care is funded, regulated, and delivered in the U. S. Some of the same kinds of distress I noted in question 2 above—for example, distress associated with the cognitive dissonance arising from increasing tension between (a) the philosophical commitments and ideals that guide hospice care as a conceptual model and (b) the increasing emphasis on financial performance and regulatory compliance that has accompanied the incorporation of hospice care into the U. S. health care system—are evident in hospice organizations. Philosophically, giving hospice organizations the tools and conceptual structures through which to understand and express that distress, as well as to assert its explanatory paradigm as one that could inform changes in the larger healthcare system, is an important and urgent task. I am engaged in this task at present, and its resemblance to the need for similar work in the philosophy of nursing is, in my view, quite strong. As such, I think I can learn from, and contribute to, parallel efforts by my colleagues immersed in the philosophy of nursing.

5. Where do you see the field of philosophy of nursing to be headed, including the prospects for progress regarding the issues you take to be most important?

I'm reluctant to put forward a guess as to where the philosophy of nursing is headed in the future (indeed, if I've learned anything from Hume, it is to be suspicious of forecasting knowledge into the future when there is so much uncertainty right now in the present). That caveat aside, I'm hopeful that the field will continue to grow. The question is helpful because I suspect such growth would benefit a great deal from prospective planning and strategic targeting of philosophical and financial resources. Were I to have a voice in that planning, I would focus on three areas.

First and foremost, philosophy of nursing should employ the tools and methods of social and political philosophy to increase the ability of nursing as a discipline to articulate its value in the civic pursuit of social goods. I have explained this more fully above.

Second, now that we have a small, but growing, literature in the philosophy of nursing it is time to begin creating an evidence base for pedagogies to teach the philosophy of nursing, especially to undergraduate nursing students. The philosophy of nursing is taught in some doctoral programs, but my sense is that a philosophy of nursing course in an undergraduate college curriculum remains a very rare find. One way to increase the depth and clarity of philosophical reflection in nursing is to recall that it is not only a *profession*, it is also a *discipline*. As such, nursing students need to not merely be taught *how* to perform this or that task. Rather, they need to be given the tools to ask and answer difficult, reflective, philosophical questions on their own. Questions like: "*Why* is performing this task the right thing to do for this patient in these circumstances?" "*How* is engaging with this patient in this manner contributing to this patient's goals of care?" "Is this pattern of interactions and staffing ratios *consistent* with the mission of the organization?" "Is it consistent with the higher goals of nursing?" "What is the relationship between what I do every day with my patients and where we strive to be, collectively, as a community?" In short, we need to look beyond board exams and accreditation requirements and ask whether or not the philosophy of nursing should have a place in the nursing curriculum that it currently does not have. The time is right to ask that question, and if we have well-developed teaching methods to prepare nurses to ask the series of questions noted above the philosophy of nursing will be strongly positioned to contribute to *educating* nurses in their *discipline* as opposed to just *training* them in their *profession*.

Finally, I think the philosophy of nursing has an important role to play in keeping nursing as a discipline honest and helping it to engage a process of ongoing, constructive, critical reflection about what values and capacities help define it as a discipline. Not unlike the Socratic gadfly, the philosophy of nursing can (and should) constantly remind nursing of what it knows and what it does *not* know. One way to do this is for those of us who work in the field to make sure we build strong, collaborative relationships with leaders in nursing–disciplinary, political, and professional leaders. And, in those relationships, when we have occasion, like Socrates, to raise questions about the wisdom of certain directions or roles for nursing within the larger healthcare and social context, we should do so. Of course, as we learn in Plato's *Apology* (Fowler [trans.] 1914), philosophers who ask such questions in pursuit of the good are not always appreciated by those to whom they pose their questions. All the more reason, I suppose, to work hard when teaching philosophy to the undergraduate nurses, thereby ensuring the next generation will take up the mantle should those of us currently in the field be thanked as Socrates was.

References

King, Imogene M. 1981. *Theory for Nursing: Systems, Concept and Process*. New York: John Wiley & Sons.

Kirk, Timothy W. 2007. "Beyond Empathy: Clinical Intimacy in Nursing Practice." *Nursing Philosophy* 8: 233-243.

Peplau, Hildegard E. 1952. *Interpersonal Relations in Nursing: A Conceptual Frame of Reference for Psychodynamic Nursing*. New York: GP Putnam & Sons.

Plato. 1914. "Apology." In *Euthyphro. Apology. Crito. Phaedo. Phaedrus*. Translated by Harold North Fowler. Cambridge, MA: Loeb Classical Library/Harvard University Press.

Sellman, Derek. 2000. "Alasdair MacIntyre and the Professional Practice of Nursing." *Nursing Philosophy* 1: 26-33.

Sellman, Derek. 2011. "Professional Values and Nursing." *Medicine, Health Care, and Philosophy* 14: 203-208.

13

Kari Martinsen

Professor

Høgskolen i Harstad og Haraldsplass Diakonale Høgskole, Bergen, Norway

---

1. How were you initially drawn to philosophical issues regarding nursing?

My journey into the world of philosophy began when I, fresh out of nursing school, was working in a psychiatric hospital in the late 1960s and early 1970s. Fundamental questions soon became urgent: how could it be that nursing science placed such a marked emphasis on abstract concepts, theoretical models and methodology, crowding out pressing existential questions? Questions regarding the fundamental human condition and the life-sustaining phenomena, such as vulnerability, sensing, mutual dependency, trust, hope, mercy, were seldom posed. And if such problems were raised, they would be considered in an abstract, rational perspective removed from real life, and not in a sensitive, receptive mode requiring expressions that were descriptive and sensitively tuned. There was little room for wondering and alternative thinking.

During this period I began studying philosophy in Norwegian universities in an attempt to approach the existential questions. However, I soon found that the way they defined this subject area chiefly involved abstract thought on life, and logical argumentation, rather than thinking based on and shaped by life as it unfolds in contexts that may be both problematic and rewarding. It was not until the later 1980s, as I became acquainted with the thinking of Knud Ejler Løgstrup (1905-1981), Danish theologian and philosopher, that a new world opened up to me. Here I encountered a down-to-earth and sensitive philosophy whose starting point was the specific situations human beings find themselves in, within which our lives already unfold or are integrated, through our sensing – with our eyes, ears, noses – into meaningful and significant relationships.

2. What, in your view, are the most interesting, important, or pressing problems in contemporary philosophy of nursing?

The most interesting and important questions to be posed by a nursing philosophy are in my opinion the metaphysical questions in the way Løgstrup handles and reflects over them. Metaphysics here is a concept encompassing questions relative to the universal and sustaining phenomena of our existence, as already noted. In essence, it denotes something about our existence which is there and which sustains existence and life without being created or produced by us. At the same time metaphysics is also a mode of thinking which enables us to deal with metaphysical questions. Thus metaphysics denotes both a way of being and a way of thinking, and this way of thinking gives us a tool for characterizing metaphysical phenomena. It is a reflection rooted in our sensing, in what makes an impression on us. In wonderment we probe into the contexts and relations unlocked by our wondering.

The most pressing problem is that philosophy is not allowed to be philosophy as such but tends to be pressed into service as a theory of science (or an ontology) for empirical science. In my view philosophy is an independent and autonomous field of study and research, separate and different from empirical science. This should not be taken to mean that the two have nothing to do with each other, because they do. Philosophy and empirical research (including clinical nursing) are independent subject fields, but they are interconnected and therefore interact in certain ways. In such an interactional perspective phenomena and thinking are linked, but they are not identical. One points to the other, separate but not independent of each other. Philosophy and empirical research (including clinical nursing) are different subject fields, but they have something to do with each other through interaction.

Philosophy is about wondering, being instinctively open, searching, looking, not goal-directed like empirical nursing and also parts of clinical nursing. Philosophy, in posing the fundamental and universal questions to our existence, may contribute to enlightening areas which have been shrouded in darkness, and critically scrutinize what has been taken for granted. In this way we may approach what has been overlooked or ignored: the most significant but also the most painful elements in our understanding of human existence.

To clinch the point made above: Philosophy cannot be understood on the basis of and via empirical research. If that were the case, their interaction and mutual exchange would be cancelled. Philosophy and empirical research can mutually amplify each other when each is accepted in its own right. Similarly, philosophy and professional clinical nursing can amplify each other, with no need to proceed via empirical research.

Both a philosophical approach and empirical research are important for professional nursing.

To approach empirical research (and clinical nursing) in a philosophical way requires philosophical training and a different way of working than what is dominant in today's empirical research, where philosophy often becomes a "superstructure" added on top of it. It follows from this that one must be allowed room for alternative modes of expression rather than being forced into a pre-set mould in order to be counted as scientific. The essay should be allowed its natural role as one scientific mode of expression among several genres. Similarly, theoretical work should not be judged on the basis of criteria valid for empirical research only. In general there should be more room for wondering, and for expressing one's findings in different ways, as demanded by the research material. In other words, the dynamic shaping of expressions is important. Not everything can, and should, be expressed in the same way.

3. What, if any, practical and/or socio-political obligations follow from studying nursing from a philosophical perspective?

Løgstrup's later philosophy on the primacy of sensing, i. e. that sensing is something we cannot get around or behind, represents a very radical sociopolitical thinking. In sensing we are receptively open to what we encounter and what makes an impression on us as receiving objects. This is diametrically opposed to the rationally invasive attitude where we as proactive agents try to gain a total understanding, and the world is merely our environment which we consider from a distance and seek to intervene in and transform in accordance with our understanding. Taking a perception-philosophical approach, on the other hand, the world is our primary beginning into which we have been integrated with our bodies and our sensing, and whose vulnerability charges us with caretaker responsibilities.

In sensing, the person is sensitively tuned, touched, and moved. Sensing is the receptability required for anything to make an impression on us. And sensing projects an autonomous communicative force: the impression is in itself pregnant with meanings, which express a resistance to our intrusions. It is essential to try to embrace a dawning understanding of the meanings carried by our impression before rushing in to impose the order of our preconceived notions. It might even be that our impression disturbs our pre-understanding, diverts it into different orbits of thinking. We are here dealing with a "conversion" in the real sense, turning from an invasive to a receiving mode. At stake here is protecting the vulnerability that our sensing creates in us. This implies a starting point diametrically opposed to the rationality of traditional philosophy. Løgstrup begins with the person's vulnerability and the person's safeguarding.

This also has implications for ethics, which takes as its starting point the existential situations people find themselves in, and not in concepts, norms, and theories. It is in the situations we encounter the vulnerability of life, the pre-ethical sustaining phenomena that we have not created, such as trust, hope, mercy, and the ethical demand to care for what is alive. Our arguments and our man-made norms and the cultures we produce are there to assist us in caring for each other in difficult situations. But cultures/norms are ambiguous, they can both constrain people and open up for trusting meetings between them. It is therefore essential not to forget the primacy of sensing, to be receptive to what encounters us. Otherwise we may infringe upon the rights of the other person by enforcing ready-made rules and norms. Ethics consists of an ethical triad – the pre-ethical sustaining phenomena (the sovereign life utterances), the ethical demand, and norms/cultures – and none of the notes of this triad can do without the other two.

4. In what ways does your work seek to contribute to philosophy of nursing?

My work may contribute to a philosophy of nursing in which the fundamental aspects of our existence, that we are vulnerable and dependent on each other and on a vulnerable universe into which we are integrated, can be seen as important. This presupposes a willingness to risk being sensitively close to life itself, and an acceptance of that sensing having a decisive impact on how we understand and express that which makes an impression on our senses.

I would like to call my philosophy the philosophy of vulnerablility, presupposing the primacy of sensing. In its brief form it can be expressed as follows: Our vulnerability points to our mutual dependency, that life is fragile and can easily be destroyed. We are imperfect humans, and our vulnerability can be turned against us and used to destroy us. But at the same time our vulnerability is a precondition for feeling empathy with the suffering of others, and an appeal for being taken care of. Vulnerability viewed in this light is a strength. It is of fundamental importance to patients and their families that the nurse is capable of taking their and her own vulnerability into account in the physical space of the health services.

Philosophy, as I understand it, is close to life itself and its vulnerability. It will contribute to the creation of perspective, depth and nuances in our understanding of situations where vulnerability is involved and at stake, but also to emending them critically. It is a philosophy that needs no detour via empirical research but can start directly from the daily experience of nurses.

5. Where do you see the field of philosophy of nursing to be headed, including the prospects for progress regarding the issues you take to be most important?

My work on a philosophy of vulnerability and sensing has found resonance in clinical nursing, but also in nursing research, not least through the dialogues I have had and still have with Katie Eriksson, the Finnish nursing science researcher. We have been conducting public dialogues for 30 years across the Nordic countries, sharing and exchanging wondering questions, which humans in essence always have been posing, the questions of charity and mercy. In 2009, as a fruit of our dialogues, Katie Eriksson and I published the book *Å se og å innse – om ulike former for evidens* (" To see and to realize – on various types of evidence"). Other important dialogue partners have been Charlotte Delmar (Denmark), Herdis Alvsvåg (Norway), Unni Lindstrøm (Finland), Patricia Benner (USA), all researchers in nursing science, as well as Tom Andersen Kjær, the Danish hospital chaplain.

As regards my study of Løgstrup's work, and my attempts to think along with and beyond Løgstrup, my conversations with Rosemarie Løgstrup, K. E. Løgstup's widow (now deceased), have been of immense value.

Translated by Vigdis Elisabeth Brekke (RN, cand. polit.)

Selected Works

Martinsen, K. 2006. *Care and Vulnerability*. Oslo: Akribe.

Martinsen, K. and K. Eiksson. 2009. *Å Se og å Innse* [To See and Realize]. Oslo: Akribe.

Martinsen, K. 2012. *Løgstrup og Sygeplejen* [Løgstrup and Nursing]. Aarhus: Forlaget Klim.

See also:

Alvsvaag, H. 2006. "Kari Martinsen: Philosophy of Caring." In *Nursing Theorists and Their Work* (6th ed) edited by A. M. Tomey and M. R. Alligood, 167-190. St. Louis, MO: Mosby Elsevier.

Lindstrøm, U. Å. 2006. "Katie Eriksson: Theory of Caritative Caring." In *Nursing Theorists and Their Work* (6th ed) edited by A. M. Tomey and M. R. Alligood, 191-223. St. Louis, MO: Mosby Elsevier.

14

Per Nortvedt

Professor

Center of Medical Ethics, Institute of Health and Society, Faculty of Medicine, University of Oslo, Oslo, Norway

---

Introduction

As a caring discipline, nursing has been rooted in the western tradition of Christian charity for almost 2000 years. Nursing has been a practical caring discipline in the European hospitals, monasteries and churches from early Christian times, following the establishing of hospitals first in the Eastern parts of the Roman empire, and later in the western parts of Europe from 600 AD (Miller 1985). This motivational and practical background of nursing also shaped many of the philosophical issues that have been most prominent in modern nursing. In particular the nature of nursing as an ethical practice in its combination of theoretical analysis with practical altruism raises many pertinent philosophical questions such as:

Is there a normative core in the pure fact of being in a relationship with the patient, i. e. in the relational phenomenology of nursing?

What is the relationship between value and knowledge in clinical nursing?

What is the relationship between theoretical and practical knowledge?

What is the identity of nursing as a science, as different from medicine and other health care disciplines?

From a philosophical point of view and with regard to many different levels of philosophical analysis, ontologically, epistemologically and ethically, nursing is a very interesting discipline.

1. How were you initially drawn to philosophical issues regarding nursing?

Nursing science does not choose to engage in philosophy because philosophy is an interesting endeavour, but because philosophy is essential for understanding nursing as a helping practice. To understand the practice of nursing means to understand the philosophy of nursing and vice versa. Essential philosophical issues spring from the nature of nursing as a relational practice. Relationships between persons, here the professional helper and a foreign person (patient) with various degrees of vulnerabilities constitute interpersonal nursing. Of course, nursing has societal and sociological, cultural and political dimensions, but it is impossible to understand the philosophical underpinnings of nursing without coming to grips with its interpersonal and relational dimension. A nurse (in most cases) stands in a relationship to a sick individual human being. What does this mean philosophically and normatively? This is a question that cannot be answered in an exhaustive way in this chapter, but it is a question which relates to all the four basic philosophical questions I initially mentioned. I will focus on these four questions in the first part of the chapter.

When it comes to the first question, the ontological question about *being-with-and-for the patient*, I will underscore two dimensions of interpersonal nursing: Relational and interpersonal nursing can be seen from a voluntaristic perspective. There are various practical obligations that a nurse chooses and feels obliged to do as part of her professional responsibilities. Choice and deliberations based upon practical competency and theoretical knowledge is essential according to the voluntaristic perspective.

On the other hand, nurse-patient relationships have a non-voluntaristic dimension. Standing in an encounter with a patient has experiential dimensions that shape the basic phenomenology of nursing. For instance, to stand in a relationship to a sick and vulnerable human being is to relate to a spontaneous and situational moral demand. Simon Critchley in his book *Infinitely Demanding* argues that: "At the basis of ethics, there has to be some experience of an approved demand, an existential affirmation that shapes my ethical subjectivity and which is the source of my motivation to act" (Critchley 2008, 23). In nursing, this demand is a moral demand of *caring-for* and *being-for* another person. It is an existential demand of relieving the suffering, easing the pain, consoling the mourning person in grief.

The essential phenomenological point here is that this demand of care is not a product of voluntary and deliberate reasoning. The moral demand is not created or invented by the nurse, and is, as a fact, not deliberately chosen by her. It is up to the nurse to respond to the appeal of care and comfort, but the caring demand it is not of her making. The demand can be resisted, it can be approved or not approved, but to contradict its existence is a failure, not of deliberation, but of perception (Jonas 1984).

With Critchley again:

> All questions of normativity, whether universalistic or relativistic, have to follow from some conception of what I am calling ethical experience. That is, without the experience of a demand to which I am prepared to bind myself, to commit myself, the whole business of morality would either not get started or would be a mere manipulation of empty formulae. (2008, 23)

My arguments so far indicate that a phenomenological understanding of nursing is essential for understanding the normative foundation of nursing. A phenomenological analysis of nurse-patient relationships does not focus on how the nurse deliberates, analyses, and reasons, but on how the nurse and the patient spontaneously experience the demands, the encounters and the relationships with patients. Focus is on immediate relational experience and not on rational analysis.

Phenomenology in fact argues that there is an affective, non-cognitive and experience-based dimension to relationships (Drabinskij 2001; Husserl 1983; Nortvedt 2008). In nursing this also raises interesting perspectives for moral philosophy. Ethical subjectivity is shaped by the experience of moral demands in the clinical encounter. For a nurse to perceive vulnerabilities, pain, and a patient's experiences of illness, is to encounter moral realities. Moral value in the clinical relationship is something that a sensitive nurse experiences, something that she sees. Moral value then, is discovered; it is not merely invented by rational capacities and is not a sole consequence of a nurse's rational deliberations (Nortvedt 2012). This takes us to the relation between value and knowledge in nursing.

To argue that value is an intrinsic and inseparable part of the phenomenology of relational nursing will raise important perspectives for a theory of value in nursing, and for our understanding of the relationship between epistemology and ethics. Given that there is an inherent moral demand in the encounter with a patient's existential reality, his or her suffering, joy, pleasure and the like, clinical realities also reveal moral realities. In a recent paper on moral realism in nursing I argue that:

> Seeing the exhaustion of respiratory imposition, a person's fear of dying, the awfulness of pain is almost never neutral, but seems to reveal distinct

> moral realities. Observing the redness of a particular wound, which often manifests an infection, does not merely signify a pathological condition. It also discloses something of ethical significance, it creates some worry, it merits some helping behavior. (Nortvedt 2012, 297)

Moral value and knowledge about the patient's clinical condition are not merely externally connected and do not represent two separate forms of rationalities. It is not only the case that one has knowledge about the patient's clinical condition, for instance the actual pathophysiology and then combines this kind of knowledge with ethical knowledge and the relevant principles to make sound ethical or moral judgments. It is also the case that human pathology and the human body are immediately and non-deliberately expressed in a value-laden language, through cues, signs, and bodily manifestations. Perception of the human body is also a perception of gestalt-like patterns that reveal moral value, be it suffering, discomfort, joy, pleasure etc. (Chappell 2009). As phenomenology seems to indicate (Husserl 1983; Jonas 1984; Levinas 1996; Loegstrup 1997), the traditional *is-ought* dictum, *pace* David Hume, must be challenged and nuanced. Paradoxically, on a descriptive phenomenological level, normative value is embedded descriptively, in the '*is*', in relational existence itself. This is to say that, when it comes to practices with an inherent normative dimension, there is a sense in which the 'ought' is always already inherent in the 'is'. This in turn is not to claim that, *contra* Hume, one can argue from an 'is' to an 'ought'. It is rather to dissolve the distinction. This has implications for the relation between theoretical and practical knowledge.

It has for nearly three decades been argued that there are different kinds of nursing knowledge, and that practical knowledge is an essential part of nursing knowledge and clinical competency altogether (Benner 1984). It is not controversial to argue that there are forms of knowledge other than theoretical knowledge that are important for nursing, and that practical-experience-based knowledge is an important form of knowledge in clinical nursing. In fact, Benner's original focus on practical knowledge in intensive care nursing and expertise has proved to be widely influential and has contributed immensely to the shaping of nursing identity and for broadening the perspective of nursing science altogether.

What has been and still is a philosophical challenge however, concerns the tacit dimension of practical knowledge. As many, including Benner, have argued, there is a dimension to practical knowledge that cannot be fully verbally articulated, but is expressed in intentions, attitudes and in practical, clinical competency. The controversial question has been on this personal and subjective character of practical knowledge and its lack of possible inter-subjective validation, control and critique.

I would agree that there certainly is knowledge in nursing that is difficult to articulate in conceptual terms and also verbally. There might be clinical cues, of which interpretation and response relies on hunches, ethical subjectivity and aspects of knowledge that cannot be fully expressed linguistically. However, this does not mean that such kinds of knowledge cannot be articulated. There is nothing mystical about the kind of tacit dimension that Benner found in clinical expertise. Most of what Benner describes as intuitions can be expressed verbally, and if not verbally, they are manifest in other kinds of articulations – aesthetically as well as in practical deliberation. So-called tacit or practical knowledge is for the most part already expressed in action and in human behaviour, it has consequences for the nurse's clinical decisions, decisions that ultimately have to be justified. And knowledge about experiences of illness and human care can be expressed in art, in novels, poetry and music, expressions of knowledge that have a long history in the philosophy of curing and healing.

An interesting question however, is how does practical and experienced based knowledge relate to normativity? To experience a moral demand in a patient's suffering, can this be explained in ordinary psychological and philosophical language? Is the moral demand comprehensible? Many have argued that to speak about moral facts and moral realities as subject independent properties is to introduce mystery into ethics. Mackie for instance, long ago argued that if there are moral properties, these properties cannot be like ordinary facts (Mackie 1990; Nortvedt 2012). Levinas and other phenomenologists on the other hand argue that normative knowledge is not knowledge in the ordinary epistemological sense (Critchley 2008). Normative knowledge inherits an intrinsic *ought*, a demand that intuitively is evident in many encounters between nurses and patients. Normative knowledge is emotional, embodied knowledge evident in the gentle touch on a body in pain, or the carefulness in a manipulation one knows must be painful to the patient. Benner did not talk about moral knowledge in this way, and to speak in this manner raises fundamental questions about the relationship between epistemology and normativity, and about the need for a deeper theory of value in nursing (Nortvedt 2012).

Still, many of the philosophical perspectives on phenomenology and epistemology presented so far are not unique to nursing. So what really are the essential philosophical issues unique to nursing? Are there any?

And what is the identity of nursing as a philosophy and a science?

Are there genuine philosophical questions that are unique to nursing or are the questions that occupy nursing philosophy merely representations of general philosophical issues that might be as relevant for other health care professions such as medicine? Can we talk about *a* nursing philosophy, or do we have to talk about philosophy in general in which there are specific philosophical issues that are important and relevant for nursing?

2. What, in your view, are the most interesting, important, or pressing problems in contemporary philosophy of nursing?

Here we have some very essential identity questions for nursing philosophy and nursing science in general. One problem for a phenomenological analysis of nursing is that both the insights as well as the problems offered by a phenomenology of nursing do not appear unique to nursing. Questions concerning intuitive knowledge, the role of affective and embodied knowledge, the question about moral demands and moral realism all pertain to relationship to patients as such and are not unique to nurse-patient relationships. Hence, the ontological questions supposedly answered by phenomenological analysis of nursing are relevant for almost all health care disciplines and for a philosophy of caring in general.

Phenomenology cannot give a substantial and exhaustive answer to what is uniquely the philosophical basis for nursing and nursing science. Rather, it is the other way around. Nursing as a practice discipline, with its relational ontology can throw light upon more general phenomenological themes, such as the role of embodiment, the relationship between fact and value, between value and epistemology, as well as questions concerning moral realism. But there is no distinct phenomenology of nursing as different from a phenomenology of medicine.

This is also the reason why many phenomenological theories of nursing (at least within the Scandinavian context, i. e. Martinsen (2000) have been accused of being too general and not unique to the discipline of nursing. However, this is only a problem insofar as the aim is to outline a unique philosophy or theory of nursing. But is this aim at all possible to attain, and if it is, is it worth pursuing?

As far as I know, attempts either by conceptual analysis, or by the creation of so-called unique nursing theories or conceptual frameworks to establish a science of nursing or a unique philosophy of nursing have not been successful. Either the conceptual frameworks have been too general or they have been too narrow to capture the nature of the discipline and practice of nursing as such. If the theoretical basis of nursing is too general, it rather captures aspects of nursing that are not unique to nursing. Some good examples are the conceptual frameworks for nursing as a science presented by Meleis and Fawcett in the 1980s and 90s and their aim to establish some overarching concepts as being unique to nursing (Fawcett 1995; Meleis 2006). But, even if proved to be essential to nursing, meta-concepts like environment, health, patient or person in no way proved to be unique to nursing.

On the other hand, conceptual frameworks that are more narrow proved to be problematic for just the opposite reason. For instance Orem's theoretical model of self-care that for years has been both influential and highly controversial in Norwegian and Scandinavian nursing science, picks out *one* specific feature of nursing as *the* relevant aspect of nursing (Orem 2001). Self-care, coping and empowerment are taken to be perspectives unique to nursing, but even more, they also aspire to explain and consequently totalize the whole domain of nursing practice. But self-care is not representative to nursing only, and even more important, nursing cannot be captured through one specific conceptual or philosophical framework. The practice, history, sociology, as well as the philosophy of nursing, are all too complex to be captured in one specific theoretical model or conceptual framework. Rather a more pluralistic and integrated approach that combines various sources of knowledge is necessary to understanding, and if possible to give guidance for nursing and its science.

Moreover, the tendency to make nursing a science based upon one specific conceptual and theoretical framework as we have seen particularly in American nursing theory is seriously problematic.

In my view, questions concerning nursing as an independent discipline and a science cannot be solved philosophically or by establishing theoretical frameworks or conceptual models. Medicine and medical science has first and foremost developed as a practical science during the last 150-200 years. There has been an evolution in research based clinical knowledge trying to understand and solve the most pressing problems of illness and disease. The evolution of medical science is not a result of philosophical endeavours, i. e. the construction of conceptual frameworks and the like, but of practical, clinical work and research. Similarly, nursing as a science will mature as a result of clinical research and by creating tools and practical knowledge and guidelines for better patient care. Philosophy can at its best clarify and question the theoretical background of the discipline and investigate some of the more basic questions that are raised from the relational and clinical background of nursing, questions introduced in the introductory part of this chapter.

However, nursing and nursing philosophy cannot only focus on the particularistic and relational aspects of nursing. Nursing is a profession with a political and societal mandate of serving people in need of health care. This means that there are pressing socio-political questions in nursing that philosophy can help to clarify and perhaps even solve.

3. What, if any, practical and /or socio-political obligations that follow from studying nursing from a philosophical perspective?

A most actual and pressing problem in nursing that philosophy can elucidate is on the question of resource allocation in health services generally and with regard to nursing particularly. If nursing philosophy shall not only seek to understand intricate epistemological and ontological problems, but make nursing practice able to function and improve in a social, clinical and political context, questions about care, proximity and justice have to be raised and debated within the community of nurses nationally and internationally. For instance, philosophy can contribute in clarifying the essential tensions in clinical nursing that follow from the increasing demands of cost-efficiency in care caused by budget constraints and increasing demands and need for health care.

A tension between good and empathic individual care with attention to particular needs and vulnerabilities, and the demands of impartial distribution of care has always existed and will continue to set increasing demands on nursing in society. Globally the situation is even more alarming, as we are watching the deaths of thousands of people, women and children that easily could be saved and cared for by means that only represent a tiny fraction of the health care expenditure and budgets in the Western world (given that health and caring resources were fairly allocated across national boundaries). So, we can ask how our moral vision is so blurred and narrow so as not to see those faraway, but mainly caring for our own (Pogge 2002). Thomas Pogge has rightly asked:

How can severe poverty of half of humankind continue despite enormous economic and technological progress and despite the enlightened moral norms and values of our heavily dominant Western civilization?

> And: Why do we citizens of the affluent Western states not find it morally troubling, at least, that a world heavily dominated by us and our values gives such very deficient and inferior starting positions and opportunities to so many people?

Nursing philosophy cannot solve this momentous ethical challenge, but it can in a more sophisticated and relevant way clarify the values involved in the context of distributing nursing resources and perhaps even give some directions of thought and practice. Thomas Nagel famously argued that one of the most pressing ethical challenges of our time is how to reconcile the demands salient to individual attention and relationships with the demands of political and institutional justice (Nagel 1988).

This problem of impartiality and partiality in care is not unrecognized by nursing ethics and philosophy, but being some of the most pressing ethical challenges of our time, perhaps nursing philosophy shall raise the agenda of national as well as global justice even stronger. Philosophically, the following questions have to be discussed:

Given that caring is relational and particularistic, what does this mean for justice, for a caring that is institutionalised and also must be impartially constrained? A nurse's moral considerations must encompass more than her own patients. She also must be aware of the consequences of her particular priorities and individualised attention for other patients with relevant needs. A nursing ethics has to balance care for particular patients with attention to the demands of distributive justice. But where are the borders to be drawn and how is a reasonable balance to be envisioned?

It is a problem for clinical nursing that relational values are constantly jeopardised and time for proper care, communication and attention to individual needs and vulnerabilities are marginalised. Nursing philosophy must deal with these problems on several levels. Nursing philosophy must address the distributive challenges from the perspective of justice and theories of justice, and from the perspective of relationships, virtues and care. Nursing philosophy can also address these issues meta-ethically. The phenomenology of clinical nursing care reveals robust moral institutions in the relational encounters between nurses and patients. Phenomenological analysis in nursing raises questions about value theory, clinical normativity, moral realism and ethical subjectivism. The question is about the normative consequences of these intuitions for practical deliberation and clinical priorities. All these are "real" and challenging philosophical areas of concern where nursing philosophy can and must be able to contribute.

4. In what ways do you think your work has contributed to the philosophy of nursing?

My own work in philosophy of nursing is related to what I take to be some of the most controversial and challenging philosophical issues with a future significance for nursing practice. One of my main interests concerns the foundation of value in nursing practice and theory. In particular I have been working on the role and nature of moral sensitivity in nursing. Papers in *Medicine Health Care and Philosophy* in 2008 as well as papers in *Nursing Philosophy* in 2006, and the *Journal of Medical Ethics* in 2008, focus on three essential aspects of moral sensitivity.

First, what is the normative foundation of moral sensitivity and what distinguishes moral sensitivity from other kind of sensitivities? Is moral sensitivity merely a question for ethical subjectivism as Hume would argue, or is moral sensitivity related to our perception of properties that exist independently of subjective evaluation and perception (moral realism). This question about moral realism in nursing is addressed in a recent paper in the *Journal of Medicine and Philosophy* (Nortvedt 2012).

Secondly, I have investigated the phenomenological basis for moral sensitivity, and try to understand its metaphysical, ontological and phenomenological basis through readings of Emmanuel Levinas and Edmund Husserl. Part of the phenomenological picture here concerns the role of affectivity and emotions in moral epistemology and in clinical perception. This is a topic that has been of my interest for nearly two decades now (Nortvedt 2008).

Finally the focus on moral sensitivity, moral realism and particularism in ethics generate an interest in the normative role of proximity and an engagement in questions concerning proximity and distance, partiality and impartiality in ethics. Questions with regard to the role of proximity in ethics as well as the normative value of relationships and partiality has been of great interest for moral philosophy since Aristotle, but has regained a renewed interest by the works of Bernard Williams, Peter Singer, and Frances Kamm. Questions about the normative value of relationships are also especially relevant to debates within an ethics of care about relational ontology, and the special contribution that this theoretical perspective can contribute to moral philosophy in general. As a result of a research grant from the Norwegian Research Council 2008-2011, two special issues concerning ethics of care and the normative role of relationships have been published in the *Journal of Health Care Analysis* (2011) and in *Nursing Ethics* (2011).

Questions about the morality of relationships and the normative value of relational obligations have a particular interest to clinical nursing because nurses constantly must be advocates of their patients, and they have to reconcile interests and needs of particular patients with demands of distributive justice and other patients with health related needs. This question about care and justice in clinical nursing has been investigated empirically in several research projects (Nortvedt, et al 2008; Pedersen et. al 2008) within the Norwegian health care context. Also a recent article has been published which discusses the so-called relational ontology of an ethics of care and the extent to which the care ethical perspective can give arguments for a certain partiality in clinical nursing care (Nortvedt, Hem and Skirbekk 2011).

5. Where do you see the field of philosophy of nursing to be headed, including the prospects for progress regarding the issues you take to be most important?

Nursing philosophy in the future should be specially engaged in philosophical discourse with high relevance for practical and clinical nursing. If nursing philosophy cannot engage substantially in issues relevant to clinical nursing, we really cannot talk about a philosophy of nursing. Nursing is ultimately for the best of patients and has a mandate from society of serving the people with regard to their health related needs. Hence, issues discussing the nature of the discipline, its normative source and the tensions between caring and curing, between care and justice must have an important place in a philosophy of nursing.

Nursing philosophy should raise the critical questions concerning the science and discipline of nursing. Philosophy of nursing should be open, engaged and an unafraid proponent of critical discourse and reflection. It should engage in investigating hidden power structures in nursing science and contribute critically, as well as constructively by way of the best argument, to current debates in nursing science and methodology.

Nursing philosophy must be curious and raise the important questions for nursing knowledge and practice. It should also aspire to integrate nursing history into the theorising of ethics and epistemology in nursing. Not least, nursing philosophy should engage and collaborate with other disciplines such as philosophy in general, the philosophy of medicine, sociology, psychology and other health care sciences.

References

Benner, P. 1984. *From Novice to Expert: Excellence and Power in Clinical Nursing Practice*. Menlo Park, CA: Addison-Wesley.

Chappell, T. 2009. *Ethics and Experience: Life Beyond Moral Theory.* London and New York: Acumen Publishing.

Critchley, S. 2008. *Infinitely Demanding – Ethics of Commitment, Politics of Resistance*. London: Verso.

Drabinskij, J. 2001. *Sensibility and Singularity- the Problem of Phenomenology in Levinas*. New York: Suny Press.

Fawcett, J. 1995. *Analysis and Evaluation of Conceptual Models in Nursing*. Philadelphia, PA: University of Pennsylvania.

Health Care Analysis. 2011. Ethics of Care, Special Issue, 19(1).

Husserl, E. 1983. *Ideas Pertaining to a Pure Phenomenology and to Phenomenological Philosophy, first book*. Dordrecht, The Netherlands:

Kluwer Academic Publishers.

Jonas, H. 1984. *The Imperative of Responsibility*. London: The University of Chicago Press.

Levinas, E. 1996. *Totality and Infinity – an Essay on Exteriority*. Dordrecht: Kluwer Academic Press.

Loegstrup, K. E. 1997. *The Ethical Demand*. London: Notre Dame Press.

Mackie, J. 1990. *Ethics. Inventing Right and Wrong*. London: Pelican Books.

Martinsen, K. 2000. *Care and Vulnerability*. Oslo: Akribe Publisher.

Meleis, A. 2006. *Theoretical Nursing*. Philadelphia, PA: Lippincott, William & Wilkins.

Miller, T. 1985. *Birth of the Hospital in the Byzantine Empire*. Baltimore: John Hopkins University Press.

Nagel, T. 1988. *Equality and Partiality*. Oxford: Oxford University.

Nortvedt, P, Pedersen R, et al (2008) Clinical Prioritisations of Health Care for the Aged – Professional Roles, *Journal of Medical Ethics,* 34: 332-335.

Nortvedt, P. 2008. "Sensibility and Clinical Understanding." *Medicine, Health Care and Philosophy* 11: 209-219.

Nortvedt, P., M. H. Hem and H. Skirbekk. 2011. "Role Obligations and Moderate Partiality in Health Care." *Nursing Ethics* 18 (2): 192-200.

Nortvedt, P. 2012. "The Normativity of Clinical Health Care – Perspectives on Moral Realism." *Journal of Medicine and Philosophy* 37(3): 295-310.

Nursing Ethics. 2011. Special Issue: Ethics of Care. 18(2).

Orem, D. 2001. *Nursing – Concepts of Practice*. St. Louis: Mosby.

Pedersen, R., P. Nortvedt et al. 2008. "In the Quest of Justice? Clinical Prioritisation in Health Care for the Aged." *Journal of Medical Ethics* 34: 230-235.

Pogge, T. 2002. *World Poverty and Human Rights*. Cambridge: Polity Press.

15

John Paley

Senior Lecturer

School of Nursing, Midwifery and Health

University of Stirling, Scotland, UK

---

1. How were you initially drawn to philosophical issues regarding nursing?

Rudely interrupted during an MSc research methods module. "Never mind all that, when are you going to tell us about phenomenology?" It was 1995, and I was teaching nurses en masse for the first time. I was completely unaware of this development in qualitative nursing research, and I was probably none too successful in camouflaging my embarrassment. I stumbled through some general remarks about the history of phenomenology, but it's difficult to believe that the students found my observations helpful. Subsequently, the embarrassment developed into curiosity, and I spent the next six months reading the nursing literature (reaction: "I'm not convinced they've got this right"), followed by the original texts – Husserl and Heidegger – and a fair chunk of the secondary philosophical literature (reaction: "Actually, I think they've got it wrong"). Given that I had done so much reading, I decided I might as well write a couple of papers. After all, shame to waste.

In the meantime, what with all the reading of the nursing literature, my attention had been drawn to a debate about intuition in the *Journal of Advanced Nursing* and to the genre of concept analysis, which was spookily like the philosophy I'd been familiar with thirty years earlier – except for the philosophy part. At all events, I ended up writing a couple of papers on concept analysis and intuition before I got round to saying something about the relation between Husserl/Heidegger and nursing phenomenology. By then, I was off and running, and building up a list of topics I wanted to write about, including more on Heidegger, dualism, caring, and paradigms.

I can't remember how I came to attend the first Philosophy of Nursing Conference, at Swansea in 1997, but I did. On the first day, I nearly fled. I was due to give a paper at a concurrent session – the planned paper on Heidegger, in fact – but I was on second, and when I entered the room the first speaker was waiting to start. Or, I should say, speakers. One was a man wearing bunny ears, another was a woman in a dressing gown, and the third was a ten year-old boy in his pyjamas. What followed was a series of mimes and tableaux which I found completely unfathomable. Worryingly, however, everybody else appeared to be enjoying the spectacle, and I began to wonder – because here I was with an erudite little number on Heidegger tucked in my back pocket – whether philosophy of nursing was something I was cut out for. Fortunately, the rest of the conference was less hard to make sense of, and I met several interesting and entertaining people. In the end, that conference and the people who attended it were instrumental in drawing me in. I was attracted to the community almost as much as I was attracted to the issues.

2. What, in your view, are the most interesting or important problems in philosophy of nursing?

I don't think there are any important problems in the philosophy of nursing, as such. But philosophical ideas, often misunderstood or distorted, have been imported wholesale into nursing research and nurse education, and the consequences are sometimes both profound and regrettable. The problem is that the nurses who do the importing often haven't got much philosophical background, so there is a risk that they will misinterpret the ideas they latch on to. Subsequently, though, they are regarded as authorities by other nurses, and the misunderstandings become credal. Decades later, there will be an established academic industry, churning out papers according to a recipe or model devised by the importing nurse, probably with some tinkerings and minor variations, but typically with few, if any, challenges to the original appropriation, and with no attempts to catch up on what academic philosophy has been doing in the meantime. The precedent set by published papers keeps the industry going in a self-sustaining cycle; for once there are several hundred examples in print, it becomes almost impossible to question the foundations of an entire approach.

I will outline three examples, commenting briefly on some wider issues they illustrate.

The first is concept analysis. The most popular version of this genre (Walker & Avant 2005) is based on a book designed to introduce high school students to analytic philosophy as practised in the 1950s (Wilson 1963). Walker and Avant cite no other sources. The kind of analysis Wilson describes sets out to determine the necessary and sufficient conditions for the application of a problematic concept. The classic analysis of 'know', for example, goes like this:

(k) X knows that p
if, and only if:
(a) X believes that p,
(b) X is justified in believing that p, and
(c) p is true.

The idea is that (a), (b) and (c) are independently necessary, and jointly sufficient, conditions for (k) to be true. However, this claim can be tested analytically. For instance, suppose we could think of a case in which conditions (a), (b) and (c) hold, but (k) does not. *Prima facie*, this would show that, while (a), (b) and (c) are necessary conditions of (k), they are not, in fact, jointly sufficient. The famous 'Gettier cases' are usually taken to show exactly this (Gettier 1963).

Two key points here. One is that cases are used to *test* the proposed analysis. The other is that empirical data is not required; the testing takes the form of *thought experiments* which show whether the analysis works or not.

Either Walker & Avant understood none of this, or else they introduced a series of significant changes without explaining why. In their procedure:

- The aim of determining the necessary and sufficient conditions for the application of a concept is replaced by the aim of specifying the concept's 'defining attributes'.
- Cases are not used to *test* the proposed analysis but to *illustrate* its results (Risjord 2009).
- The methodological gap vacated by cases is occupied by the frequency criterion: the 'defining attributes' are those which appear 'over and over again'. Thought experiments, in other words, give way to a literature search and a bit of counting.
- The 'antecedents and consequences' of the concept are introduced, though they are irrelevant to a philosophical analysis (and Wilson does not mention them).

The result of the first two changes is that sufficient conditions are likely to get left out. If you are just looking for 'attributes' of a concept, you will find yourself listing (at best) necessary conditions; and, since cases are no longer used to test the analysis, it will be impossible to determine whether the listed 'attributes' are jointly sufficient or not. The upshot is that, in practice, many concept analyses in the nursing literature do no more than specify a few necessary conditions. They certainly do not 'define'.

For example, an analysis by Cowles and Rodgers (2000) delivers the 'definition' of a particular concept, as follows: 'a dynamic, pervasive, highly individualized process with a strong normative component' (108). Any idea what that is? I have asked several students and colleagues to work out what is being defined here. None of them has got even close (remember, this is presented as a *definition*). It might be almost anything. No, I'm not going to tell you. Look it up if you want to know. Then ask yourself what theoretical, clinical or research use such a 'definition' could possibly have.

The other two changes make things even worse. The frequency criterion implies that, from a list of 'things said' in the literature, the most popular ones will be selected as 'defining attributes'. But there is, of course, no guarantee that frequently mentioned characteristics are even necessary conditions, let alone jointly sufficient ones (Paley 1996). The problem is compounded by the fact that antecedents and consequences are often included among the 'defining attributes'. Walker and Avant (2005) started this tradition early. In their 'attachment' example (78-9), they confuse antecedents with defining attributes, despite warning against this very confusion earlier in the chapter (73).

So this type of concept analysis in nursing amounts to a popularity index of 'things said' in the literature. These 'things said' are neither necessary nor sufficient, and they often include causes and effects of X, rather than the conditions for describing something as '*X*'. The genre is, in fact, a mess. It produces either arbitrary lists of necessary conditions – which do not define, and which are often banal – or uncritical literature reviews tortured into a 'defining attributes' framework that just does not fit.

In any case, developments in philosophy and psychology make the Walker and Avant method look badly outdated. For example, the idea that the structure of a concept can be identified with its 'definition', or with its necessary and sufficient conditions, has been almost universally rejected. In its place, a number of theories have been discussed and tested: the prototype theory, the exemplar theory, the causal theory, the originalist theory, and conceptual atomism (Williamson 2007, Sainsbury & Tye 2012, Baz 2012). Conceptual analysis as a project has been challenged by: philosophical naturalism, which subordinates analysis to empirical data (Papineau 1993); a renewed interest in ordinary language philosophy, which rejects the claim that concepts have an invariant internal structure in the first place (Wilson 2006, Baz 2012); and concept eliminativism, which suggests that 'concept' denotes too many different kinds of thing (Machery 2009). Yet concept analysis in nursing sticks largely to its Walker and Avant groove.

Nurses who wish to continue this tradition might find McGinn (2012) useful. McGinn outlines a form of analytic philosophy not far removed from Wilson's; but anybody reading his book should note that the position he adopts is wildly unfashionable. A recent review lists numerous authors and arguments that he simply ignores (Ahlstrom-Vij 2012).

For those who prefer to do something more in keeping with the philosophy and linguistics of the last 50 years, there are a number of options (since what counts as 'concept analysis' will vary with the purpose of the enquiry). One is to consider semantic questions of 'lumping and splitting', as lexicographers call it (Kilgarriff 2008). For example, why is it fashionable to 'lump' together situations involving a patient with gambling debts, a retired teacher missing the job, a vegan refusing hospital meals containing dairy produce, and a patient not wanting a gastronomy tube inserted (McSherry 2006) with problems having to do with religion, in the concept of 'spiritual need? Why not keep them 'split' – that is, differentiated? Another is to assess the performative function of certain terms, instead of assuming that they all have 'descriptive' meaning (Baz 2012). The semantic 'lumping' implicated in the concept of 'spiritual need' has a clear ideological function (Paley 2008, 2010). So, arguably, does the term 'caring', which cannot be said to possess descriptive content (Paley 2002). Instead, it is part of the 'nursing 'brand'. It has the same function as the phrase 'Because you're worth it'. It is the nursing profession's tagline. Just as Nokia has been 'Connecting people' since 1992, so nursing has been 'Indomitably caring since 1860'.

There are plenty of alternatives. Discriminating between semantics and pragmatics (Wilson and Sperber 2012); the formulation of stipulative definitions (Lycan 1994); studying the relation between empirical data and decisions about classification (LaPorte 2009); the use of corpus linguistics to identify textual contexts and collocations for critical terms (Stubbs 2002); and so on. All of these draw on recent work in linguistics and philosophy, and are considerably more useful than identifying 'attributes'.

My second example is the topic that introduced me to philosophical issues in nursing. The belief that the phenomenology of Husserl and Heidegger points towards a method for qualitative research is found in several disciplines – particularly psychology and education – so, in this case, it is not just nurses who have got hold of the wrong end of the stick. Indeed, it was perfectly reasonable for nurses to assume, 30 years ago, that methods texts by authors like van Manen (1990) and Giorgi (2009) were authoritative guides. It is just unfortunate that they did not have the philosophical acumen to recognise these authors' limitations and to interrogate their work properly.

At any rate, the general assumption that the focus of phenomenology, for both Heidegger and Husserl, was 'experience' appears to be ineradicable, though it is completely mistaken. Husserl wanted to break out of experience (into the realm of pure consciousness) through the phenomenological reduction; and Heidegger explicitly disowns the concept – along with 'subject', 'consciousness', 'person', and 'spirit' – because it belongs to the dualism between 'experiencing subject' and 'experienced object' that he is trying to dismantle (Heidegger 1962, 72). Being-in-the-world has absolutely nothing to do with lived experience (Keller 1999); neither does Husserl's pure transcendental consciousness (Paley 1997, 1998).

Equally wrong is the view that, according to Heidegger and Husserl, 'essences' or 'essential structures' can be identified by collecting and analysing empirical data. For both of them, identifying essences is a wholly philosophical enterprise. It is a form of *a priori* analysis, and empirical data is simply irrelevant. Yet the conviction that the 'essence of a phenomenon' can be investigated by interviewing people, and looking for 'common themes', persists.

This might not matter so much if it were not for the fact that a great deal of so-called phenomenological research in nursing is poor: trite, contradictory, and clinically valueless. Much of this is a consequence of a philosophical misunderstanding – the view that, according to phenomenology, 'reality *consists* of the meanings in a person's lived experience' (Omery and Mack 1995), a view which neither Husserl nor Heidegger ever subscribed to – combined with the conclusion that there is no distinction between 'how things are' and 'how things seem' (Paley 2005b). This leads naturally to the idea that, in order to study *anything*, it is necessary only to ask people a few simple questions about their 'experience' of it.

In turn, the consequence of *that* view is a painting-by-numbers approach, based on a kind of pro forma research design. For if enquiry is by definition the exploration of somebody's experience of something, then it follows that any study can be defined by filling in the blanks of a one-size-fits-all research aim:

> To { explore / illuminate / elucidate / gain insight into [1] } … *<description>*
>
> the{ experience / meaning / meaning of the experience [1] } …*<experience>*
>
> of { *insert concept, situation, treatment or condition* } …*<concept>*
>
> from the perspective of { nurses / patients /health

professionals [1] }<*perspective*>

in{ *insert location* }<*location*>

[1] *Delete whichever does not apply*

This is an intellectually undemanding exercise, which takes the pain out of thinking about research; and the only price to be paid is a willingness to compose a few garbled sentences about ontology, linked to one's author of choice: Husserl, Heidegger, Gadamer, Merleau-Ponty, Ricoeur, or whoever.

It is important to recognise that nursing phenomenology and concept analysis share some underlying philosophical assumptions. Both of them imply a sort of homogeneous ontology in which experiences, events, behaviour, conditions, situations, practices, accomplishments, projects, judgments, decisions, perceptions, relationships, meanings, values, motives, emotions, beliefs, and other psychological states are flattened into a single ontological category – 'concept' or 'phenomenon' – and studied in the same way. Walker and Avant (2005), for example, do not consistently recognise the distinction between the *what-is-classified* (event, condition, project) and the *what-does-the-classifying* (concept). Phenomenologists in nursing blur the distinction between the *phenomenon* and the *experience-of-the-phenomenon*.

Moreover, every type of thing in this homogeneous ontological category has *attributes*, in deference to the quasi-Aristotelian metaphysics that runs through nursing theory like a philosophical watermark. It is this metaphysics which lurks beneath a great deal of nursing research, and which leads to findings which are effectively lists of attributes: lists of features, lists of characteristics, lists of themes, lists of factors, lists of barriers... in short, and in Aristotelian terms, lists of 'things said' about the phenomenon being studied. Devising these lists of 'things said' is construed as identifying the attributes of an event, activity, situation, condition, experience, phenomenon, concept, or absence. In nursing, it sometimes seems that the only type of thing that counts as knowledge is a list of attributes (Paley 2001).

This flattened-out metaphysics fails to be properly Aristotelian, despite a superficial similarity, because it confuses 'essential attributes' with 'accidental attributes'. The list of 'things said' regularly fails to distinguish between attributes that are essential to X (attributes whose absence would entail that the object in question was not an X at all), and those that are merely 'accidental' (the object would still be an X even if these attributes were not present). For example, it is an essential attribute of water that it is a liquid at normal temperatures and pressures;

it is an accidental attribute that it comes out of a tap.

In concept analysis, the confusion between 'essential' and 'accidental' is a result, as I observed earlier, of Walker and Avant's 'frequency' criterion for identifying 'defining attributes'. If you look for features that 'are mentioned over and over again', you are far more likely to identify 'common-but-accidental' attributes than you are to identify essential ones. The situation is comparable in phenomenology. The search for 'common themes', incorporated into the 'essential structure' of the phenomenon, reflects the assumption that essences can be determined empirically, and is responsible for the confusion between 'essential' and 'common-but-accidental'. In both cases, bad philosophy makes it difficult to distinguish between the *nature* of an event, situation, or condition; the way in which it is *experienced*; and the way in which it is *conceptualised*. The consequence is uninformative, why-did-anyone-bother research.

Phenomenology and concept analysis are both entrenched in the nursing literature for several reasons. One, as I have suggested, is that recent philosophical debate – which, in effect, means the debate of the last 50 years – rarely percolates into nursing. Another is that both genres are convenient exercises for postgraduate students: small-scale, self-contained, intellectually undemanding, recipe-dominated and, in the case of concept analysis, not requiring research ethics approval. This is why they have survived repeated criticism on both methodological and philosophical grounds.

The third example is knowledge or, as nursing often prefers to say, 'knowing'. The literature has seen a great deal of debate on this topic, particularly since the advent of evidence-based practice, much of it characterised by the continued veneration of Carper's seminal paper, from 1978, and an unwillingness to review developments in epistemology and the philosophy of science since the 1960s.

Like authors in other disciplines (a recent example is Moses and Knutsen 2012), many nurses appear to think that the philosophy of science stopped after Kuhn and Feyerabend. Although there are exceptions, they are generally unfamiliar with more recent debates about the significance of models in scientific practice, the focus on mechanisms rather than laws, constructive empiricism, Bayesian conceptions of scientific inference, perspectivism, realism and antirealism, essentialism and natural kinds, the interest in cognitive science, computer models of scientific inference, the new experimentalism, the return of reductionism, increasing attention to scientific values, the study of large-scale collaboration in science, historical analyses of the different conceptions of objectivity, the exact role of mathematics in scientific theories, causal modelling, and the emphasis on philosophy of particular sciences, especially biology, physics, and neuroscience. Nor is there any recognition of how work on the philosophy of language in the 1970s and 1980s, after Kripke, took the ontological and epistemological sting out of worries about Kuhnian incommensurability and its supposedly relativistic implications.

One consequence of this is the persistence of Carper's claim that there are other 'patterns of knowing'. Nursing has been rather reluctant to give up the idea that several 'ways of knowing' have comparable epistemological status, different but equal; and there has been no serious discussion of the reasons why science can, and should, be privileged relative to other so-called 'patterns'. If, for example, scientific methods are distinguished by the fact that they incorporate procedures for identifying and reducing the risk of inferential error (Mayo 1996), then we must ask whether other 'patterns of knowing' incorporate procedures with the same function. If they do not, that is a good *prima facie* argument for claiming that science always takes precedence (Paley 2005a). Recent work in psychology, indicating that there are at least two neurally differentiated cognitive systems, one of which is abstract, explicit, logical, and rule-based, reinforces this argument, especially in view of the fact that the rule-based system has an audit and supervisory function with respect to the other systems (Evans and Frankish 2009). So the claim that equal status should be accorded to 'other ways of knowing' implies that methods that can identify and minimise error should carry no more weight than methods that can't. In effect, it celebrates the inability to recognise error (Paley *et al.* 2007).

This applies not only to Carper's 'patterns', but to other kinds of non-scientific 'knowing'. Consider, for example, the claim that 'knowing how' and 'knowing that' are different types of knowledge. This idea has been popular since Benner (1984), and implies that 'knowing how' is cognitively distinct from, and epistemologically independent of, 'knowing that'. But this view can be maintained only by ignoring work in philosophy which argues that 'knowing how' is *reducible* to 'knowing that'. It is possible to establish, on linguistic grounds, that 'knowing how' falls into the category of 'knowing-WH' constructions, which also includes knowing-what, knowing-where, knowing-whether, knowing-why, and so on. All these constructions involve knowing the answer to a question, and this translates into knowing-that, a knowledge of the facts. For a review of the philosophical literature and a defence of this position, see Stanley (2011).

Similar arguments could be mounted against the other 'ways of knowing' which the nursing literature finds congenial, including 'embodied knowledge', 'experiential knowledge', 'intuitive knowledge', 'tacit knowledge' and 'professional craft knowledge'. All of these pretend to something which they do not possess: namely, an epistemological status which transcends habit, assumption, speculation, belief, conviction, reflex, faith or prejudice (Paley 2006). If they were described instead as 'ways of coming to believe', there would be no problem, and it would be possible to enquire into the validity and reliability of the beliefs concerned. But the claim that 'tacit knowledge', or any of the other examples, is a form of *knowledge* begs the question, because it builds validity and reliability into the definition of a cognitive *state of mind*. For anything to count as 'knowledge', there must be a normative promise, an offer, or a presumption, of truth; and this is precisely what these 'ways of knowing' fail to provide.

Developments in philosophy, psychology and linguistics undermine nursing's preferred epistemology. But, equally, they suggest alternatives. When nurses get round to jettisoning the ideas encrusted with age and misunderstanding, they will discover a more exciting, more varied, and more clinically useful range of philosophical resources.

3. What, if any, obligations follow from studying nursing from a philosophical perspective?

No obligations. It's the ideas – the opportunity to read, think, talk, write – that grab me. To that extent, I'm basically a hedonist. Still, I do think that nursing would benefit from getting rid of the dead wood, the misconceptions, the garbled interpretations. So, for that reason, it's worth writing about them.

Teaching is different. One of my aspirations is to inoculate my postgraduate nursing students against the gobbledegook that assails them in journals. Part of the reason why bad philosophy has prevailed for so long is the intimidation of students by authors who try to convince their readers, and I suspect themselves, of their intellectual credentials with polysyllables and prolixity. After all, 'cotranscending with the possibles is powering unique ways of originating in the process of transforming' (Parse 2007). The response of most students to this sort of thing is to imagine that they are stupid because they don't understand it. It's almost a form of abuse, and I have little respect for the writers who do it. So I try to compensate, encouraging students to ask the simple and obvious questions... and to be deeply sceptical when they don't get clear answers. Shifting them from 'It has been published, so it must be okay, and I must be an idiot' to 'I am entitled to expect clarity, and I have a right to be suspicious of authors who don't provide it' is (when it happens) one of my most satisfying achievements.

It's probably worth adding that 'scepticism when you don't get clear answers' applies to my own work as much as to anybody else's. If I can't explain a line of thought to someone else, especially a student, then I don't understand it myself.

4. In what ways does your work seek to contribute to philosophy of nursing?

I don't think it does. The 'philosophy of nursing' is not something I contribute to or try to develop. My main interest is in studying the impact of philosophical ideas – very often misunderstood, distorted, or mangled – on nursing research, and indirectly on nurse education and nursing practice, and in showing why these ideas have had a debilitating effect. I am aware that the message is not a congenial one.

5. Where do you see the field of philosophy of nursing to be headed?

Pass. I am much more interested in where nursing, as an academic discipline and a clinical profession, is headed – or at least those facets of nursing that are affected by the bad and/or out-dated philosophy I have been describing. A couple of years ago, I attended a research conference at which some of the speakers extolled 'nursing epistemologies', 'nursing's ways of knowing', and 'methods unique to nursing'. Many of the people attending lapped it up, but others fidgeted their scepticism. My fear, as far as the future of nursing and nursing research is concerned, is that the lappers will outnumber the fidgets. My hope is that I will turn out to be wrong."

References

Ahlstrom-Vij, K. 2012. "Review of 'Truth by Analysis: Games, Names, and Philosophy.'", by Colin McGinn. *Notre Dame Philosophical Reviews*. 2012. 06. 06 at:
http://ndpr. nd. edu/news/
31222-truth-by-analysis-games-names-and-philosophy/.

Baz, A. 2012 *When Words Are Called For: A Defense of Ordinary Language Philosophy*. Cambridge, MA: Harvard University Press.

Benner, P. 1984. *From Novice to Expert: Excellence and Power in Clinical Nursing Practice*. Menlo Park, CA: Addison-Wesley.

Cowles, K. V. and B. L. Rodgers. 2000. "The Concept of Grief: An Evolutionary Perspective. In *Concept Development in Nursing: Foundations, Techniques, and Applications, 2nd ed* edited by B. L. Rodgers and K. A. Knafl. Philadelphia PN: Saunders.

Evans, J. S. B. T. and K. Frankish. eds. 2009. *In Two Minds: Dual Processes and Beyond,* Oxford: Oxford University Press.

Gettier, E. L. 1963. "Is Justified True Belief Knowledge?" *Analysis* 23: 121-123.

Giorgi, A. 2009. *The Descriptive Phenomenological Method in*

Psychology: A Modified Husserlian Approach. Pittsburgh: Duquesne University Press.

Heidegger, M. 1962. *Being and Time*. Oxford: Basil Blackwell.

Keller, P. 1999. *Husserl and Heidegger on Human Experience*. Cambridge, UK: Cambridge University Press.

Kilgarriff, A. 2008. "I Don't Believe in Word Senses." In *Practical Lexicography* edited by T. Fontenelle, 135-151. Oxford: Oxford University Press, Oxford.

LaPorte, J. 2009. *Natural Kinds and Conceptual Change*. Cambridge, UK: Cambridge University Press.

Lycan, W. G. 1994. *Modality and Meaning*. Dordrecht: Kluwer.

Machery, E. 2009. *Doing Without Concepts*. New York: Oxford University Press.

Mayo, D. G. 1996. *Error and the Growth of Experimental Knowledge*. Chicago: University of Chicago Press.

McGinn, C. 2012. *Truth By Analysis: Games, Names, and Philosophy*. New York: Oxford University Press.

McSherry, W. 2006. *Making Sense of Spirituality in Nursing and Health Care Practice: An Interactive Approach*. London: Jessica Kingsley Publishers.

Moses, J. W. and T. L. Knutsen, T. 2012. *Ways of Knowing: Competing Methodologies in Social and Political Research, 2nd ed*. Basingstoke, UK: Palgrave Macmillan.

Omery, A. and C. Mack. 1995. "Phenomenology and Science." In *In Search of Nursing Science* edited by A. Omery, C. E. Kasper and G. G. Page, 139-158. Thousand Oaks, CA: Sage.

Paley, J.

1996."How Not To Clarify Concepts in Nursing." *Journal of Advanced Nursing* 24: 572-578.

1998."Husserl, Phenomenology and Nursing." *Journal of Advanced Nursing* 26: 187-193.

1998."Misinterpretive Phenomenology: Heidegger, Ontology and Nursing." *Journal of Advanced Nursing* 27: 817-824.

2001."An Archaeology of Caring Knowledge." *Journal of Advanced Nursing* 36(2): 188-198.

2002)"Caring as a Slave Morality: Nietzschean Themes in Nursing Ethics." *Journal of Advanced Nursing* 40(1): 25-35.

2005a."Error and Objectivity: Cognitive Illusions and Qualitative Research." *Nursing Philosophy* 6: 196-209.

2005b."Phenomenology as Rhetoric." *Nursing Inquiry* 12(2): 106-116.

2006."Evidence and Expertise." *Nursing Inquiry,* 13(2), 82-93.

2008."Spirituality and Secularization: Nursing and the Sociology of Religion." *Journal of Clinical Nursing* 17: 175-186.

2010."Spirituality and Reductionism: Three Replies." *Nursing Philosophy* 11(3): 178-190.

Paley, J., H. Cheyne, L. Dalgleish, E. A. S. Duncan, and C. A. Niven. 2007. "Nursing's Ways of Knowing and Dual Process Theories of Cognition." *Journal of Advanced Nursing* 60: 692-701.

Papineau, D. 1993. *Philosophical Naturalism*. Oxford: Blackwell.

Parse, R. R. 2007. "The Human Becoming School of Thought in 2050." *Nursing Science Quarterly* 20: 308-311.

Risjord, M. 2009. "Rethinking Concept Analysis." *Journal of Advanced Nursing* 65(3): 684-691.

Sainsbury, R. M. and M. Tye. 2012. *Seven Puzzles of Thought and How to Solve Them: An Originalist Theory of Concepts*. New York: Oxford University Press.

Stanley, J. 2011. *Know How.* New York: Oxford University Press.

Stubbs, M. 2002. *Words and Phrases: Corpus Studies of Lexical Semantics*. Oxford: Blackwell Publishing.

van Manen, M. 1990. *Researching Lived Experience: Human Science for an Action Sensitive Pedagogy.* Albany, NY: State University of New York Press.

Walker, L. O. and K. C. Avant. 2005. *Strategies for Theory Construction in Nursing, 4th ed.* Upper Saddle River, NJ: Pearson / Prentice Hall.

Williamson, T. 2007. *The Philosophy of Philosophy.* Oxford: Blackwell.

Wilson, D. and D. Sperber. 2012. *Meaning and Relevance*. Cambridge, UK: Cambridge University Press.

Wilson, J. 1963. *Thinking With Concepts*. Cambridge, UK: Cambridge University Press.

Wilson, M. 2006. *Wandering Significance: An Essay on Conceptual Behaviour.* Oxford: Clarendon Press.

16

Mary Ellen Purkis

Dean, Faculty of Human and Social Development, Professor, School of Nursing

University of Victoria, Victoria, Canada

---

1. How were you initially drawn to philosophical issues regarding nursing?

My earliest introduction to philosophical issues, although I am not certain that I recognized that was what I was being introduced to at the time, was in my undergraduate nursing program. The school I attended had adopted the work of Sister Callista Roy (Roy 2009) as a model around which to organize the nursing curriculum. Reading Roy's model of adaptation represented a quite separate domain of experience than what I recall when I would engage in the range of practice settings we were introduced to as part of the nursing program. I had chosen to enter a university to study nursing rather than to attend the hospital school of nursing to which I had also applied and gained admission. Reading about nursing in the rather abstract and highly systematized terms that it was presented through Roy's text was oddly reassuring to a novice nurse who otherwise found my exposures to nursing practice on hospital wards chaotic and deeply concerning. Patients' conditions could change so suddenly and the expectations that I had set for myself to respond to those changes made the clinical setting one that I approached with considerable anxiety! Also, I was very conscious in those early days that I had made a very deliberate choice in attending the university rather than the hospital-based school of nursing and the abstract nature of the reading felt like the right sort of alignment with my choice of educational programs.

Roy's adaptation model offered a way for me to organize my thoughts about what I understood I was expected to do as a nurse. It assisted me to impose an order on myself and on my work with patients. It was particularly useful to me during my time as a student nurse. I have taken opportunities to reflect on that experience (Purkis 2007) and have previously noted that, remarkably, it did not seem to matter that none of the nurses I worked with knew about nor used Roy's model to guide their practice. Just as it served as a way of buttressing my choice of educational programs, it also served to insulate me from the world of everyday nursing work. It afforded me an "academic" space within which to engage with that foreign yet exciting world. It assisted me in getting my assignments done, in planning my day as a student assigned to a clinical unit and in accounting for my actions when asked by some – but not all – of my nursing instructors. The model was most strongly embraced by teachers in the early part of the four-year program – less so by those who taught in the final year or two. By the time I reached the fourth year of my program, I was being challenged by my teachers to think beyond Roy's model, to draw on a wider range of intellectual and practical resources to support my practice.

That was very good advice on a number of fronts.

Following graduation from my nursing degree program, I worked for two years in a busy, downtown emergency department. That experience enabled me to hone my knowledge and skills in assessing and making good judgments about people's suffering bodies and minds. As I came to trust my own judgments and the support of my colleagues in recognizing and treating these illnesses, I found that I relied less and less on the organizational structure that Roy's model had provided me as a student learning about patterned ways of providing nursing care.

After those two years, I traveled to the UK and practiced there for nearly two years before enrolling in a graduate degree at Edinburgh University. That degree led to a qualification in nursing education. It was in preparing my teaching plans as a novice teacher that I found myself falling back again into Roy's model to help me organize my thoughts and prioritize what I would share with student nurses, initially in the UK and later back in Canada when I was hired to teach in a two-year college program in southern Alberta.

It was perhaps being in this position as a teacher that I began truly to engage with the philosophy of ideas that constitute contemporary nursing. During those three years of my first teaching appointment I experienced the limitations of my existing academic preparation for the task of gaining a satisfactory position *on* nursing. I had relied pretty successfully on Callista Roy's framework to get me through my undergraduate degree program, to make the move into a busy, acute care practice environment, to put myself in the uncomfortable situation of transporting my Canadian nursing background into the new health care delivery context of the United Kingdom and then to serve as the disciplinary substance to support my new practice as a nurse educator.

But the discourse of nursing was changing through the 1980s and when I began my career as a nurse educator I found myself being challenged to provide an account of nursing as somehow contributing to the promotion of health – not just to the assessment of disease and the establishment of plans and interventions designed to treat those diseases. And in this conceptual space, I was dissatisfied with what was available to me when I did what I had by now become accustomed to doing: turning back to my old, familiar conceptual framework provided by Callista Roy. I had always been able to use Roy's model to break big problems down into more manageable component parts whether I was assessing someone with chest pain in the emergency room or planning a class on pain management for Scottish nursing students. The promotion of health seemed to call for something more holistic and Roy's atomistic model no longer applied. I began searching for new ways of organizing ideas about nursing that could help me solve this problem of avoiding disease and breaking the experience of being healthy down into component parts.

An influential text uncovered during this period of searching was Judith Smith's article entitled, *"The idea of health: a philosophical inquiry"* (Smith 1981). Smith's work opened up many possibilities for me: that there was a field of scholarship in nursing that relied on methods other than those of natural science, that concepts such as "health" and even "nursing" were contested and unsettled, and maybe most importantly, that I did not need to keep *looking back* to find answers to my questions but that I could *look forward* to fields of scholarship that I now had a sense of – but was not yet introduced to. It was with this latter hope that I made the decision to return to Edinburgh for doctoral studies – and there, I was fortunate to meet colleagues who introduced me to forms of inquiry that have continued to satisfy my interests in looking forward in an effort to describe and explain the practices of nursing.

2. What, in your view, are the most interesting, important, or pressing problems in contemporary philosophy of nursing?

The most pressing problem facing contemporary philosophy of nursing is the lack of a convincing body of literature that builds a productive bridge between the s(t)olid body of writing generated primarily in the United States of America from the late 1950s through to the early 1980s that has taken on a framing as "the nursing theory movement" and the, dare I say, interesting, studies of nursing that have found a supportive home in journals such as *Advances in Nursing Science* (in the early years of that journal), *Nursing Inquiry* and *Nursing Philosophy* – particularly over the last two decades. Much of this latter writing represents a counter discourse to the earlier theoretical texts – but the ways in which it positions itself in contrast to that very particular theorizing of nursing is rarely explicated. Because of this, we do not have an *effective* philosophical bridge over which the history of ideas in nursing can be conveyed and, most importantly, engaged critically.

Some may argue that this is simply a by-product of geography: there is no doubt that the theory movement was much more a feature of the North American nursing academy than it was in the United Kingdom, Europe and Australia. But this argument I think does not satisfy. Some very influential British philosophical groups have built a substantial body of philosophical scholarship by engaging in critical dialogue with the American theorists. What this critique has been most effective at drawing our attention to is the narrow base upon which nursing scholarship has been founded. What it has not done as effectively, in my view, is to offer us explanations for why that narrow base was so attractive to the early theorists, what the effects on nursing scholarship have been our largely unquestioned reliance on that early theory movement for the development of our discipline and then to begin to sketch out what the boundaries of the discipline could be and which theoretical and philosophical perspectives might best enable us to explore those boundaries. It may be that an important outcome of this text and all those who contribute to it, is that we may finally bring those many voices into one physical space where these questions can be engaged and new programs of philosophical research may begin.

3. What, if any, practical and/or socio-political obligations follow from studying nursing from a philosophical perspective?

The most important, entirely practical obligation arising out of a philosophical engagement within the discipline of nursing is, for me, the demand to explore the complex relationship between the nurse and the suffering body. It is *in that space* that nursing is accomplished and therefore, it is *that space* that demands philosophical attention. I would like to offer just one example of such scholarship in order to illustrate my point here.

Per Nortvedt holds a professorial position in the Centre for Medical Ethics at the University of Oslo. He has drawn on the work of Emmanual Levinas to examine such topics as "sensibility" as it pertains to clinical understanding (Nortvedt 2008). In this paper, Nortvedt focuses on those moments when a patient's suffering is noticed by the nurse and where such suffering is experienced by the nurse as a "sensory-based distress felt by the distress of another human being" (210). This is a deeply philosophical concern – but one that is equally practical. If the nurse fails to notice the other, fails to experience distress in the face of the other's distress, the nurse cannot respond to the other. Here is an example of a philosophical engagement in that complex relationship operating between nurses and patients.

In this paper, Nortvedt asks the reader to consider the implications of an overly influential perspective in moral philosophy of emotions as defined by cognitive features, definable, describable, organized and organizing of scholarship that seeks to tame the emotions and put them to good and careful purposes in the development of health care professionals. This has, no doubt, been an important corrective to earlier prescriptions of good, professional practice that advocated cool objectivity as the preferred stance of the professional caregiver in relation to the patient. But Nortvedt argues that focusing on the cognitive aspects of emotions over-plays an interest in judgment, while undermining the "central and epistemologically (...) more crucial and independent role" (214) of the affective part of the nurse's emotional sensibility in response to the patient's suffering.

While Nortvedt does not tackle "the theory movement" in nursing directly, in focusing on the dominance of cognitive discourses within literature supporting the development of the professional *persona* of health care providers including nurses, his work stands as an example of philosophical writing that engages productively with that pressing problem that I note above. Nortvedt's paper lays out an argument for considering the clinician as an emotional subject where "emotions are part of responding, not merely at the right time, with reference to the right objects and toward the right people [all of which serves as the substance of a cognitive approach to professional ethics], ... but responding in the right manner and with the right attitude" (Nortvedt 2008, 213). He goes on to state that, to arrive at a clinical judgment about the extent to which a patient suffers, but not to be moved by that suffering is, in effect, "not to see what the suffering really amounts to" (213). Here, Nortvedt's contribution opens up an opportunity to examine practice as a socio-politically informed phenomenon. Drawing on a sophisticated reading of Levinas' idea of ethical sensibility, he sets out the possibility that the clinician and the patient meet one another in the space of nursing practice and in recognition of their radical otherness; that is, in a way that the clinician can only ever attempt, but never be entirely successful, in bridging the difference between the two beings.

So here we have a remarkable "moment" within which nursing practices and the relationships that obtain between the nurse and the patient's suffering body can be investigated: Why does the nurse respond to the call of the patient? What has the nurse seen when she glances at the patient as she passes his bed? What is it that she sees that enables her to drop the other calls on her time in order to seek further information from the patient who suffers so as to formulate an always provisional response? And perhaps more importantly, what does the nurse see when she turns away from a patient who calls for her attention? How does the nurse enact herself as a nurse, as a member of a team of caregivers, as an employee of a health authority when the calls on her time supercede her capacity to respond? According to Nortvedt's reading of Levinas, the nurse is in an on-going process of self-definition every moment she works and moves towards some instances of suffering and away from other instances of suffering. She cannot *not* respond. "Levinas tries to capture the condition of the subject in which it is vulnerable and passive in receptivity towards the other" (216). This is not a philosophical perspective that demands adherence from practitioners – rather, in Nortvedt's rendering of Levinas, it is *descriptive* of their encounters with patients. The nurse subject is vulnerable and passive in her receptivity of the suffering of the patient. What she then goes on to do about her sensibility of the suffering of the other in terms of how it informs her clinical understanding and from there, her clinical actions, can be traced empirically.

Such philosophical work can inform sophisticated studies of nursing practice as active, political, influential and consequential for the well-being of patients.

4. In what ways does your work seek to contribute to philosophy of nursing?

By the time I left my initial teaching position in order to begin my doctoral program, I had begun to read more widely and, as noted earlier, focusing on what was quite a vexing problem: how to conceive of the practice of nursing as promoting health? When I first came across this question, I was teaching young student nurses registered in a two-year college program. The efforts of teachers contributing to this program were aimed at ensuring the students were competent practitioners upon graduation – and the measure of their competence was, without exception, based on evaluations of their practice in acute care hospital environments. It was not just the teachers who were concerned with the competence of the soon-to-graduate students: the students themselves also held this concern. A two-year program offers little room for exploration of the breadth of practice opportunities for nurses when both students and teachers are focused on a clear goal of ensuring graduates of the program are "job ready." And so, pity the poor instructor whose job it was to teach these students about health promotion in their final term prior to graduation! The students were distracted and difficult to convince that the topic would be relevant for their up-coming examinations. I was discouraged by the nature of the literature on the topic. My recollection of the literature was that it reflected three broad themes: (1) proclamations from national and provincial nursing bodies that health promotion was the rightful terrain of professional nursing; (2) literature suggesting health promotion properly described the practice of community health nurses; and (3) literature emanating from international bodies such as the World Health Organization where nursing was rarely, if ever mentioned in association with health promotion. Instead, health promotion was proposed as a new and necessary approach to achieve sustainable forms of health care for populations around the globe.

Based on my untutored reading of nursing philosophy at the time, I felt myself persuaded primarily by authors whose research was based in qualitative frameworks and perhaps mostly by writing that claimed association with phenomenology. These texts offered a rich, if not entirely comprehensible, language describing nursing in highly positive terms. There was a certain mystical quality of these relationships between nurses and their patients who were said to have "lived experiences" of this and that – but always something that nurses could positively affect. I was attracted by these ineffable accounts of nursing practice – all the more so as I struggled to put into words this new form of practice known as health promotion.

It was with all these questions in my mind that I began my doctoral work. Despite my attraction to phenomenology, I found as I was pushed by other students as well as potential supervisors of my research to formulate my question, I ended up in a very practical stance in relation to this troublesome topic of health promotion. What I really wanted to know was "How do nurses do health promotion?"

I was remarkably fortunate to have registered in a Philosophy of Science course in my first year as a doctoral student taught by a philosopher and social theorist by the name of Rolland Munro. In this class I was introduced to a wide range of philosophical texts by someone who was fluent with the range of perspectives. Rolland was the first professor I encountered in my doctoral program who did not seek to dissuade me from asking a "how" question. Indeed, he invited such questions – and he offered a wealth of resources to me to help me find an answer to my question. I was guided into a study of this question through a reading of works by philosophers such as Hans-Georg Gadamer (Gadamer 2004) and Alfred Schutz (Schutz, Embree et al. 2011) and social theorists such as Anthony Giddens (Giddens 1984), Erving Goffman (Goffman 1959), Harold Garfinkel (Garfinkel 1967) and, through all of this, Michel Foucault (Foucault 1979; Gordon and Foucault 1980).

Drawing this reading together, I set off to try to find answers to my question about how nurses *do* health promotion in their practice. And what I discovered is that within nursing's writings on its own practices, we have not developed ways of conceiving of practice as a dynamic and social phenomenon (Purkis 1994). Two particular texts were important to me in developing my ideas about practice. The first was Peggy-Anne

Field's ethnography of public health nursing practice (Field 1980) and the second was Patricia Benner's influential book, *From Novice to Expert* (Benner 1984). At the time I was engaged in my doctoral research, ethnographic studies of nursing practice were hard to come by – and so Field's ethnography provided an extremely useful contrast to my own ethnography of public health nursing practice. Our ethnographies were undertaken approximately 10 years apart – and from vastly different philosophical perspectives: mine a Foucaultian-inspired archaeology, hers by George Mead's social psychology. It was not only the coincidence of a mutual interest in the work of public health nurses that attracted me to Field's work but also the premise upon which she engaged in her study: "If one accepts the assumption that the core of any profession lies in its practice, then to understand that profession it is necessary to study practice within the contextual setting" (Field 1983, 3). If I wanted to discover how nurses did health promotion, I seemed to be on the right track! Ultimately I found Field's interpretations underdeveloped – largely because her theoretical foundation had no way of accounting for power in the relationship between the nurse and the patient. I turned to Benner's (Benner 1984) work here and was impressed by her commitment to the study of practice – something remarkably few nursing scholars have done. But again, I found Benner's work similarly silent in terms of its ability to account for power in the relationship between nurses and patients, and this, despite the fact that her most influential text, *From Novice to Expert*, is sub-titled "excellence and power in clinical nursing practice."

It was only with the benefit of a now closely tutored reading of Foucault, Giddens, Garfinkel and Goffman that I was able to generate a quite different reading of the health promoting practices of public health nurses and to offer an interpretation of that practice as one that, while quite coercive, could be conducted in a highly efficient and "light" manner – largely at a considerable distance from where the practice occurred in what was known as the "well baby clinic." Nurses offered parents small "tips" on how to care for their child during the clinic visit and these tips were observed to frequently occasion substantial alterations to the physical set-up of their homes as well as their ways of interacting with their child when they returned home following their clinic visit – ostensibly for regular immunization.

In this work and what has followed, I seek to theorize nursing as a social accomplishment – by which I mean that to practice nursing means that one is engaged in a political intervention, that nursing actions always have impact and that the interpretations of those impacts cannot be pre-determined but are, instead, contingent on circumstances within which the practice is enacted.

I have taken this approach into studies of nursing practice within the context of home care, noticing the ways in which standard features of hospital nursing can apparently so easily and without regard to what the individual or family may want, be replicated in the home, turning it, for all intents and purposes, from a home into a hospital (Purkis 2001; Purkis and Bjornsdottir 2006). And from those early home care studies, now into studies of nursing practice with older adults living in community and residential care settings (Ceci, Björnsdóttir et al. 2012).

5. Where do you see the field of philosophy of nursing to be headed, including the prospects for progress regarding the issues you take to be most important?

As I consider this question I am aware of what I see as a tension arising from that literature that demonstrates a cleaving to the origins of nursing theory development as a way of legitimating nursing as a particular form of women's work and, against this, the still unfocused desire to present a romanticized version of a form of work that, with reference to forms of phenomenology, are produced in ways that seem so distant from the grim 'realities' of today's nursing workplaces. Neither of these tendencies will, in my view, support nursing to fulfill its historic promise to "care for the sick" (Nelson 2001).

The longer I work in academic nursing, the more generous I feel I am growing in terms of supporting a broad range of approaches to scholarship in this important field of caring practices. At the same time, I feel my focus narrowing onto scholarship that can appreciate and demonstrate the impact of what Nortvedt describes as "sensibility" and how such sensibilities towards the suffering of the other can be shown to influence clinical judgment. We are entering another era of cost containment in health care and at these times the knowledge that nurses bring to their practice is frequently disregarded in favour of staffing arrangements that can be said to provide "coverage" of the physical space within which patients suffer. Those interested in contributing to a contemporary philosophy of nursing need not concern themselves with legitimating the practice of nursing as an academic practice, nor convince readers of the general "good" of the practice: rather, we must focus on the ways that a professional nurse grasps the seriousness of the suffering of her patients and how she turns that sensibility of suffering into a knowledgeable response that is *qualitatively different* – and I suspect *qualitatively better* – than those who are being hired to replace her.

References

Benner, P. 1984. *From Novice to Expert: Excellence and Power in Clinical Nursing Practice*. Menlo Park, CA: Addison-Wesley Pub. Co., Nursing Division.

Ceci, C., K. Bjornsdóttir, and M. E. Purkis. 2012. *Perspectives on Care at Home for Older People*. New York: Routledge.

Field, P. A. 1980. *An Ethnography: Four Nurses' Perspectives of Nursing in a Community*. Edmonton, AB, University of Alberta. Dissertation.

Field, P. A. 1983. "An Ethnography: Four Public Health Nurses' Perspectives of Nursing." *Journal of Advanced Nursing* 8(1): 3-12.

Foucault, M. 1979. *Discipline and Punish: the Birth of the Prison*. New York: Vintage Books.

Gadamer, H.-G. 2004. *Philosophical Hermeneutics*. Berkeley: University of California Press.

Garfinkel, H. 1967. *Studies in Ethnomethodology*. Englewood Cliffs, NJ: Prentice-Hall.

Giddens, A. 1984. *The Constitution of Society: Introduction of the Theory of Structuration*. Cambridge: Polity Press.

Goffman, E. 1959. *The Presentation of Self in Everyday Life*. Garden City, NY: Doubleday.

Gordon, C. and M. Foucault. 1980. *Power/Knowledge: Selected Interviews and Other Writings, 1972-1977*. New York: Pantheon Books.

Nelson, S. 2001. *Say Little, Do Much: Nurses, Nuns, and Hospitals in the Nineteenth Century*. Philadelphia: University of Pennsylvania Press.

Nortvedt, P. 2008. "Sensibility and Clinical Understanding." *Medicine, Health Care and Philosophy* 11(2): 209-219.

Purkis, M. E. 1994. "Entering the Field: Intrusions of the Social and its Exclusion from Studies of Nursing Practice." *International Journal of Nursing Studies* 31(4): 315-336.

Purkis, M. E. 2001. "Managing Home Nursing Care: Visibility, Accountability and Exclusion." *Nursing Inquiry* 8(3): 141-150.

Purkis, M. E. 2007. "Aftermath of the Curriculum Revolution: What Did We Overthrow?" In *Teaching Nursing: Developing a Student Centered Learning Environment* edited by L. Young and B. Patterson, 366-384. Philadelphia: Lippincott, Williams and Wilkins.

Purkis, M. E. and K. Björnsdóttir. 2006. “Intelligent Nursing: Accounting for Knowledge as Action in Practice.” *Nursing Philosophy* 7(4): 247-256.

Roy, C. 2009. *The Roy Adaptation Model.* Upper Saddle River NJ: Pearson Prentice Hall.

Schutz, A., L. E. Embree, et al. 2011. *Collected Papers.* New York: Springer.

Smith, J. A. 1981. “The Idea of Health: A Philosophical Inquiry.” *Advances in Nursing Science* 3(3): 43-50

17

Mark Risjord

Professor of Philosophy, Emory College of Arts and Sciences
Emory University, Atlanta, Georgia, USA

---

1. How were you initially drawn to philosophical issues regarding nursing?

The Nell Hodgson Woodruff School of Nursing began its PhD program in 1999. The faculty consulted with me about the creation of a core course covering issues in nursing theory and the philosophy of science. I quickly realized that I knew too little about nursing research to advise them adequately. Philosophy of science should not speak *ex cathedra*; the sciences provide both content and relevance for philosophical problems. The fastest way to get up to speed was to help teach the course, refining the syllabus as we went along. I agreed, and I have been part of the course for every year since.

Like many American PhD programs in nursing, the philosophy of science is a required part of the Emory curriculum. Typically, the students have had a rich background in nursing practice, but little experience with research. They are suddenly thrown into an environment where they must critique scientific literature, formulate questions, and choose methodologies. The core questions of the philosophy of science—questions like "What is theory and how is it tested?" or "What does it mean to be scientific?"—have a personal dimension for these young scholars. As I began to understand the nursing literature, I came to see that the students' journey from nurse to nurse scholar mirrors that taken by the nursing discipline (especially in the US) during the 20th century. What had been abstract, even arcane, philosophical questions about theory structure and confirmation suddenly took on a new significance. Ideas from philosophy, both bad ones and good ones, had already shaped the nursing discipline, and they were continuing to shape nurse scholars. When I recognized the direct impact that philosophy was having on nurse education and practice, I knew I had to get involved.

As I moved from teacher to author in the philosophy of nursing, the pedagogical motivation remained important. I have often chosen topics that would be important for young nurse scholars as they formulate their research projects. My first publications in this field (Risjord, Dunbar and Moloney 2002; Risjord, Moloney and Dunbar 2001) engaged the then-febrile debate over whether qualitative and quantitative methods could be mixed. The arguments in the literature drew on claims about theory structure and about the ontological presuppositions of theories and methods. In this area, the literature in the philosophy of science could be brought to bear on a question of immediate importance for nurse researchers. At the same time, the philosophy of science in nursing is not a simple application of philosophical knowledge to a new area. The questions that begin their lives as problems for investigators grow into philosophical questions in their own right, and can have deep influence on the way we understand knowledge, ontology, or ethics.

2. What, in your view, are the most interesting, important, or pressing problems in contemporary philosophy of nursing?

The philosophy of nursing is a multi-faceted enterprise. One should therefore refrain from identifying any problems as *the* most important for the entire field. I will restrict my remarks to those philosophical questions that arise out of nursing research and scholarship, what might be called the philosophy of nursing science (understanding "science" in broad and inclusive terms, and most emphatically including qualitative research).

Nursing has a tradition of attention to the philosophy of science that goes back to the nineteen fifties. As nursing schools in the United States became more closely affiliated with universities, research became a more prominent part of the discipline. Nurse scholars found that they needed a framework for nursing research. That is, they needed something that would show how and why nursing research was distinctive and what unified it into a coherent discipline. Nurse scholars looked to philosophers of science to help address these questions. The philosophy of science and nursing changed together through the nineteen sixties and seventies. By the early nineteen eighties, a consensus developed within the community of (again, predominantly American) nurse scholars who wrote about the character of nursing research and the nursing discipline. Nursing, it was argued, was unified by the concepts contained in a "metaparadigm" (person, health, environment, and nursing). Within this conceptual orientation, there were multiple paradigms. Abstract and general—so-called "grand"—theories were to provide orientation for middle-range theories. Ultimately, these nursing theories were to guide nursing practice.

The dominant framework for the disciplinary knowledge of nursing which was established in the nineteen eighties has become both practically and philosophically problematic. In *Nursing Knowledge* (Risjord 2010) I argued that the often-decried relevance gap between theory and practice arose and persisted because of the philosophical framework for nursing research. Both nursing practice and nursing scholarship have outrun the philosophical understanding of how theory and practice fit together. The most important issues in the philosophy of nursing science, therefore, arise out of the need to reconcile our philosophical understanding of nursing research with the real demands of nursing practice and nursing scholarship.

A fundamental issue concerns the unity of the nursing discipline as a scholarly enterprise. What, if anything, is distinctive about *nursing* research? The consensus of the nineteen eighties held that the concepts of the nursing metaparadigm and the grand theories of nursing unified the discipline. Since the eighties, the influence of grand theory has waned; few research articles today bother to fit their questions or results into a theoretical framework unique to nursing. Most nurse researchers would describe themselves as developing middle-range theory. (That we continue to speak of a "middle" without commitment to a "top" illustrates the continuing power of this philosophical framework.) The concepts of the metaparadigm have remained, in the words of John Paley, "inert, like four garden gnomes" (Paley 2006, 277). If the grand theories cannot link middle range theories together in a substantive way, the problem of unity becomes sharper. Is nursing research really nothing but a hodgepodge of theories and empirical results developed by faculties of nursing schools? If not, then what links them together into a discipline?

The problem of unity thus has epistemological and political dimensions. Epistemologically, the issues concern how theories, models, methods, and research programs fit together. Should we think of them as packaged into more or less discrete paradigms? Must there be higher-level theories from which middle range theories are deduced? Or is science just a loosely related collection of models and methods? A commitment to any of these views of science has consequences for the way that nursing fits into the academy. Emphasizing grand theory and the hierarchical conception of scientific theories supported arguments that the practice of nursing should be buttressed by an academic discipline. If science is merely a patchwork of overlapping models, then perhaps there is no distinctive nursing discipline. With a patchwork understanding of science, how then do nurses argue for the importance of the nursing PhD, or for university-based nurse training? And without a unified knowledge base, how can nurses defend the importance and uniqueness of their contribution to health care? In a practice discipline like nursing, epistemological questions about nursing knowledge will always be linked to political questions about nursing's social mandate. For this reason, the problems associated with understanding the unity of the field are some of the deepest questions of the philosophy of nursing science.

Nursing research and scholarship often needs to integrate domains that are elsewhere treated separately. A typical nursing problem will involve physiological, psychological, social, and moral dimensions. One of the reasons that nursing should not be considered an "applied science," it has often been argued, is that knowledge drawn from one field must be transformed in the light of the other aspects of the nursing problem. This gives rise to a family of philosophical questions about how particular concepts, theories, and methods are to be related. The great debate over whether "qualitative" and "quantitative" methods could be mixed, blended, or triangulated is one of these. Or again, many concepts that are central to nursing practice, like *pain* or *stress,* are two-sided in the sense that they have both an experiential side (pain hurts) and a physiological side (pain is the activation of a system of nerves). These raise questions that touch both on the philosophy of science and on the philosophy of mind. Are psychological predicates or statements reducible to biological ones? What are the consequences of reductionism for nursing practice? If psychological and biological theories must remain distinct, how do we understand these inter-level concepts that are so important to nursing practice? How can theories of pain or stress be tested? These issues have taken on additional importance as genetics, neuroscience, and cellular biology have become part of nursing knowledge and practice.

3. What, if any, practical and/or socio-political obligations follow from studying nursing from a philosophical perspective?

Closing the theory-practice gap is perhaps the most important practical and socio-political consequence of studying the philosophy of nursing science. The point of having a scholarly arm of nursing is, presumably, to improve nursing practice. Unfortunately, many nurses today, those on the floor as well as those in the academy, feel that current nursing research does not speak to the problems of practice. Nursing theory is regarded as too abstract and vague, and attempts to make nursing practice "evidence based" often feel alien. In *Nursing Knowledge,* I argued that the relevance gap has arisen, at least in part, from a philosophical framework which emphasized grand theory and demanded vertical integration of theory from the most abstract to the most practical. This led to research programs that were disengaged from practice, in spite of the researchers' best intentions. It also led to the isolation of the nursing discipline, cutting it off from the useful results of other disciplines. A better understanding of nursing research and scholarship ought to highlight the connection between theory and practice in a way that makes relevance a natural outcome of nursing research.

4. In what ways does your work seek to contribute to philosophy of nursing?

In my work, I argue that the scope of nursing practice should determine the domain of nursing scholarship. The key idea is drawn from feminist standpoint epistemology. On this view, when a social role is oppressed, there is knowledge available from that role's perspective which is not available from other perspectives. On the classic analyses, *e. g.* Smith (1974) or Hartsock (1983), while oppressed groups must be able to understand the social world as the dominant groups do, they also have access to the workings of that world in a way that is not available to the members of the dominant group. The housekeeper who empties the trash and cleans the sheets is in a position to know about the social relations of her employers in a way that is largely invisible to them. Standpoint epistemology argues that unearthing the knowledge available from a standpoint takes both a political commitment to justice for or valorization of the social role, as well as real epistemic work. Epistemic privilege is not something that arises automatically from an oppressed social position. Indeed, oppression typically makes its own conditions invisible, even to those who suffer from it. Because the conditions of oppression are hidden, it takes committed investigation to expose them.

Standpoint epistemology has taken race, gender, and class to be the main dimensions of analysis, and it has treated social relationships as the object of knowledge. My analysis has extended the argument to the social roles within health care, and I take the object of knowledge to be human health. The social mandate of nursing puts nurses in a position that is structurally identical to the groups on which standpoint epistemologists have focused their analyses (and, of course, the dimensions of gender, race, and class are directly relevant as well). Nurses have to be able to work in the physician-dominated world of the biomedical sciences, but at the same time, they must understand a realm of health phenomena largely invisible to the physicians. The nursing standpoint captures the uniqueness of nursing knowledge.

Understood as a standpoint, the disciplinary knowledge of nursing must be grounded in nursing practice because it is only through nursing practice that the relevant phenomena can become visible. Research and scholarship must take up the nursing problems and subject them to serious analysis and rigorous investigation. No special methods or theories are needed to unify the discipline, and nurse scholars are free to draw on other disciplines. Theories and models are integrated into a coherent body of knowledge by addressing questions that arise from nursing practice. Treating nursing knowledge as a standpoint thus answers the question of unity in a way that encourages scholars to borrow theories and models of phenomena that have been developed and tested in other fields, and to develop them by integrating them into multifaceted answers to nursing questions.

Regarding nursing research and scholarship as arising from the nursing standpoint has a much better chance of closing the relevance gap than the prevalent top-down philosophical frameworks. On the top-down view, concepts of the metaparadigm and grand theory determine the direction of nursing research. There is no guarantee, then, that nursing knowledge will be relevant to nursing practice. From the perspective of the nursing standpoint, inquiry becomes *nursing* research insofar as it engages questions related to the problems faced by practicing nurses. Proper nursing research is therefore always relevant. Of course, the relevance may be indirect. It would be a mistake to infer that since nursing knowledge arises from the nursing standpoint, all research and scholarship must be intervention or outcome oriented. Making nursing practice better is the ultimate goal, but many forms of research and scholarship are needed. Some research questions may be very abstract, even philosophical, and relatively remote from practice. They are tied to nursing practice insofar as they related back via chains of questions and answers. There are very real gaps between theory and practice, but these should not be problems of relevance. Rather, they are problems of translation, education, and institutional change.

In my view, the problem of unity is solved by the scope of nursing practice, not by a special object of study or unique theoretical posit. Resolving the problem of unity in this way has (at least) three consequences. First, it changes the status of theory borrowed from other disciplines. The use of borrowed theory has long been problematic to nurse scholars. Drawing on established disciplines, rather than developing theories unique to nursing, seemed to dilute the discipline or turn it into an applied science. On the view articulated in *Nursing Knowledge,* relying on established theories strengthens the field, rather than weakening it. Epistemologically, the claims of a theory are linked as question to answer. Striking phenomena give rise to questions, and scientific models and theories answer them. These theories give rise to further questions, and other theories and models provide the answers. The large-scale epistemology of science, then, is an *explanatory coherence* epistemology (see Risjord 2000 for development of this idea with special reference to qualitative research). On this view, good scientific theorizing requires linking to other domains, including domains at the same level of abstraction. The links are, so to speak, horizontal as well as vertical. *Nursing Knowledge* thus argues that borrowing is not only possible, it is necessary. Only by drawing on explanations from other disciplines can nursing knowledge be epistemologically robust.

The second consequence concerns the way nurse scholars understand concept development and concept analysis. In a tradition that goes back to the nineteen seventies, nurses have thought of concept analysis as something that happens at the beginning of an inquiry. Faced with new phenomena, nurses must first create new concepts, then link them into propositions (Walker and Avant 2005). Drawing on well-established views in the philosophy of language, Paley (1996) argued that this notion of concept analysis is incoherent. Concepts have no content outside of their function within larger bodies of discourse, and in science, this discourse is theory. Concept development is part and parcel of theory development. In "Rethinking Concept Analysis" (2009), I developed the consequence of Paley's critique that, if concept analysis is to have any value at all, it needs to articulate a concept either as it already appeared in theories, or as a kind of qualitative research. In the former case, what I call "theoretical concept analysis," its value is to see disjunctures among the ways in which a phenomenon is conceptualized by different fields, disciplines, or theories. Understanding these differences in the way that a concept like stress, for example, functions in different theories is an important first step in modifying those theories so that they explain phenomena important for nursing. In the second kind of case, what I call "colloquial" concept analysis, the object is to identify the meaning of a concept as it is ordinarily used by a particular group of speakers. This is a kind of a qualitative research that is well illustrated by ethnography and ordinary language philosophy.

Third, grounding the unity of nursing in a source outside of theory and theory structure permits a different approach to the evaluation of theory and method. One of the most difficult problems confronting young scholars is to decide which theories and methods are the most useful for their projects. Existing criteria (e. g. Fawcett 2005) emphasize the relationship to grand theory and present formal criteria for evaluation such as simplicity, consistency, and clarity. Formal criteria tell a scholar nothing about the content, and presumably the content is the basis for relevance, not form. If we see theories and models as not a hierarchy but as a patchwork of overlapping explanatory systems, more helpful criteria emerge. A theory is potentially useful, on this view if it answers (at least in part) the nurse scholar's questions. The question then is whether the theory (or model) is the *best* of the alternatives. Theories should be compared, I argue, on the following grounds:

1) Operationalizability: the extent to which the elements of the theory/model can be linked to observation and, if appropriate, measured.

2) Precision: the extent to which predictions can be made with the theory/model.

3) Empirical support: the existence of successful direct tests of the theory/model.

4) Theoretical integration: the extent to which the theory has explanatory links to other theories/models at both the same level of abstraction and those at higher or lower levels.

5) Idealization: whether the theory ignores, abstracts away from, or idealizes in ways that matter for our understanding of the nursing phenomenon.

6) Values: whether the values encoded or presupposed by the theory/model are defensible and consistent with other values of the research.

Again, these criteria are comparative. The question for a nurse scientist is whether the theory under consideration is the best explanation of the nursing phenomenon with which she is concerned.

Finally, *Nursing Knowledge* argues that while methods and theories are interdependent, they are not nearly as tightly linked as many nurse scholars have supposed. It is common for nursing authors—especially those defending qualitative research—to contend that qualitative and quantitative research form paradigms within nursing. A paradigm, according to Kuhn (1962), includes a theory and its ontological commitments, methods used to support the theory, and value commitments. Philosophers of science since Kuhn have argued compellingly that domains of inquiry are not incommensurable. While methods require theoretical justification, the theories which support the method should be distinct from the theories which the method tests. Methods can therefore travel among scientific domains. It follows that methods need to be chosen for their informational, epistemic value. The scholar needs to consider how the method will interact with the object of study (whether this is a person, a social system, or some part of the human body), and how those interactions will provide information that answers questions drawn from the theory.

The main contribution of my work in nursing philosophy, then, is to critique the philosophical commitments that have been present in the nursing literature about nursing science. By sweeping away some of the bad philosophical ideas that still litter the literature, we can build a new way of understanding the field of nursing science. It is my goal to develop these new philosophical ideas into practical guidance for nurse scholars who are developing and testing theories that address the problems of nursing practice. In this way, I hope, nursing philosophy will help close the theory-practice gap.

5. Where do you see the field of philosophy of nursing to be headed, including the prospects for progress regarding the issues you take to be most important?

With respect to the philosophy of nursing science, I see two important challenges for the future. First, 20th century science made awe-inspiring advances in our understanding of the micro-mechanisms that stand behind biological, psychological, and social phenomena. From the mechanisms of gene-environment interaction, to neuroscience, to the cognitive foundations of social interaction, these advances are producing results that are relevant to nursing practice. They are increasingly becoming part of nurse education and nursing practice. Both nursing practice and nursing research, however, have resisted incorporating such advances on the grounds that they are "reductionist." Reliance on micro-mechanisms conflicts with the holistic character of nursing practice, where phenomena need to be addressed in their complexity. Practicing nurses cannot resolve their problems by separating biological, psychological, and social dimensions of a patient. The interesting philosophical challenges in the 21st century, then, revolve around this conflict between the holism required by practice and the reductionism suggested by theory.

As we dig deeper into these issues, I think we will find that some of our fundamental concepts are changing. Pain, for example, was at one time treated as an entirely experiential, psychological phenomenon. If there is an ontological gulf between minds and bodies, then pain is on the side of the mind, not the body. Yet it is becoming common sense to regard pain as a phenomenon of the nervous system. While it remains striking, it is starting to make sense to say that an unconscious patient is in pain. Our concept of pain is thus undergoing a rapid transformation—the kind, I think, envisioned by Richard Rorty in *Philosophy and the Mirror of Nature* (Rorty 1979). Nursing philosophy can and should be on the forefront of that conceptual change.

References

Fawcett, Jacqueline. 2005. "Criteria for Evaluation of Theory." *Nursing Science Quarterly 18*(2): 131-135.

Hartsock, Nancy C. 1983. "The Feminist Standpoint: Developing the Ground for a Specifically Feminist Historical Materialism." In

Discovering Reality edited by Sandra Harding and Merrill B. Hintikka, 283-310. Dordrecht: D. Ridel Publishing Company.

Kuhn, Thomas. 1962. *The Structure of Scientific Revolutions*. Chicago: University of Chicago Press.

Paley, John. 1996. "How Not to Clarify Concepts in Nursing." *Journal of Advanced Nursing 24*: 572-578.

Paley, John. 2006. "Book Review: Nursing Theorists and Their Work, Sixth Edition." *Nursing Philosophy 7*: 275-280.

Risjord, Mark. 2000. *Woodcutters and Witchcraft: Rationality and Interpretive Change in the Social Sciences*. Albany, NY: SUNY Press.

Risjord, Mark. 2009. "Rethinking Concept Analysis." *Journal of Advanced Nursing 65*(3): 684-691.

Risjord, Mark. 2010. *Nursing Knowledge: Science, Practice, and Philosophy*. London: Wiley-Blackwell.

Risjord, Mark, Sandra Dunbar and Margret Moloney. 2002. "A New Foundation for Methodological Triangulation." *Journal of Nursing Scholarship 34*(3): 269-275.

Risjord, Mark, Margret Moloney and Sandra Dunbar. 2001. "Methodological Triangulation in Nursing Research." *Philosophy of the Social Sciences 31*(1): 40-59.

Rorty, Richard. 1979. *Philosophy and the Mirror of Nature*. Princeton: Princeton University Press.

Smith, Dorothy E. 1974. "Women's Perspective as a Radical Critique of Sociology." *Sociological Inquiry 44*(1): 7-13.

Walker, Lorraine Olszewski and Kay Coalson Avant. 2005. *Strategies for Theory Construction in Nursing* (4th ed.). Upper Saddle River, NJ: Pearson Prentice Hall.

18

Gary Rolfe

Professor of Nursing

University of Swansea, Swansea, Wales, UK

Introduction

This is an opportune moment to be thinking not only about the philosophy of nursing but about philosophy as a discipline and a way of thinking. Alain Badiou (2011), in his *Second Manifesto for Philosophy*, observed that whereas twenty years ago philosophy was in danger of disappearing, it is nowadays under threat for the diametrically opposed reason that it is everywhere, and that its intellectual currency has become devalued to the point of worthlessness. Badiou's prescription for philosophy is that it must once again become *reckless* or else lose its soul completely. By 'reckless', I understand Badiou to mean that philosophers should not be overly concerned with the consequences of their actions; that they should place values above outcomes, means before ends.

1. How were you initially drawn to philosophical issues regarding nursing?

As someone who studied philosophy before becoming a nurse, a more appropriate question might be: *'How were you initially drawn to nursing?'*, to which I can only offer the perennial answer given by philosophy graduates down through the ages: *'because I needed a job'*. The rigours of studying for my final examinations left me determined never to read another book, and in the three years between finishing my degree and starting my nurse training I read little or no philosophy. As a mental health nursing student, however, I reluctantly began to read again, and developed an interest in the work of R. D. Laing and the so-called anti-psychiatry movement. Although I had read (and failed to understand) Laing's short book *The Politics of Experience* whilst at university, it was his earlier seminal work *The Divided Self* that later reignited my interest in philosophy.

Laing was a self-taught philosopher who began as a schoolboy by reading Freud, Kierkegaard and Nietzsche, whose books he discovered by working his way from A-Z through the Govan Hill Public Library in Glasgow. He later moved on to Hegel, Jaspers and Buber, and claimed in an interview that he learned French by reading Sartre's *L'Être et le Néant* and then went on to learn German in a similar way by attempting to translate Heidegger's *Sein und Zeit*. He explained that he looked up every word as he went along for the first ten pages, after which he had grasped the basic vocabulary and it thenceforth all became very easy! By the time he came to publish *The Divided Self* in 1959, Laing had a working knowledge of all the major continental existentialist philosophers and a particular interest in Dilthey and the hermeneutic tradition.

My own education in philosophy had been more traditional and probably less extensive than Laing's, but I had taken a course on existentialism and written my dissertation on Sartre. When I came to read the introductory chapters to *The Divided Self* I therefore had a little insight into what I took to be Laing's fundamental proposition that phenomenology and hermeneutics offer the best hope of understanding the lives and experiences of others. Laing famously applied this approach to his work on schizophrenia, which he claimed is not a medical condition or an illness, but is rather a way of coping with life. As such, it is best understood ontologically rather than biologically or chemically. Hence, we should not think of a person as 'having' schizophrenia in the way that they have heart disease or influenza, but rather of adopting a schizoid or schizophrenic *way of being*. My own understanding of Laing led me to believe that this was a valid and authentic way of approaching not only psychiatry but psychiatric nursing: indeed, Laing's rejection of psychiatric medicine as anything more than symptom management offered nurses an opportunity to play a major role – *the* major role – in the understanding and treatment (in the wider sense of the word) of patients presenting with what Laing called 'ontological insecurity'.

As my career progressed and I became involved in the full range of activities expected of a practising nurse (and later, of a researcher and an academic), I was drawn to a variety of philosophical 'issues', mostly concerned with ethics (what should we do?), epistemology (what can we know?) and the philosophy of science (how can we be sure that we know it?). Later, as I became (pre)occupied with writing for publication, I turned to Barthes, Derrida and the poststructuralists for an understanding of writing and its relationship to thought and action. Although my early philosophical education has helped me enormously in thinking about these issues, my work has mostly taken the form of philosophy *for* nursing rather than philosophy *of* nursing. I would argue that whilst a broad philosophical perspective is useful to nursing, and whilst ideas and principles from philosophical schools and individual philosophers are often brought to bear on nursing issues and problems, this does not in itself amount to a coherent philosophy of nursing, nor yet even the beginnings of a general programme of philosophical research. If philosophy of nursing is ever to achieve the disciplinary status of the philosophy of science or even the philosophy of education, then we must move beyond simply applying (say) principles of Aristotelian virtue ethics or a knowledge of Cartesian dualism to pre-existing nursing problems. This, in my opinion, is a poor kind of philosophy, hardly philosophy at all. If it is indeed the case that philosophy can be applied to problems from other disciplines (that is to say, whether terms such as 'philosophy of science' have any substantive meaning) and that nursing is one of those disciplines that it can meaningfully be applied to (unlike, say, geography), then we are only just taking the first steps on a very long and arduous academic journey.

2. What, in your view, are the most interesting, important or pressing problems in contemporary philosophy of nursing?

I am mindful when attempting to answer this question that I do not approach it as a question of politics. It would be tempting as a philosopher to use this question as an opportunity to map out a curriculum or to stake out some territory within the academic discipline of nursing that belongs to philosophy and therefore to philosophers; subject areas or curriculum content that is protected or ring fenced from other nurse academics who have no training or educational qualifications in philosophy. From this 'land grab' perspective, the problems for philosophers of nursing might include almost any problem of medical or nursing ethics, issues related to 'philosophical' research methodologies such as phenomenology and hermeneutics, and wider epistemological questions of the nature and production of nursing knowledge(s). However extensive the range of issues, however long the list of topics, this nevertheless amounts to little more than a programme of philosophy *for* nursing; of the colonisation of the discipline of nursing by an occupying force.

Having called into question the existence and perhaps even the very possibility of philosophy of nursing as an integrated and substantive field of study, I would like briefly to think a little more about philosophy and its relationship with other disciplines before speculating on what might be the most pressing problems for nursing. Whilst academic nursing is a relatively new arrival to academia, philosophy is arguably its founding discipline. Writing at the time of the birth of the modern University, Kant was anxious to ensure that philosophy retained its rightful place in the academy, although this place was not as might be expected, as the highest of the faculties and departments, but as the 'lower faculty'. In *The Conflict of the Faculties*, Kant divided the university into three 'higher faculties' and a 'lower faculty'. The higher faculties are so called because their teaching is sponsored and directed by the government and their chief function is to serve the public good. Kant identified these faculties as theology, law and medicine, devoted respectively to the eternal, civil and physical well-being of the citizens. However, he argued that the university also requires a faculty that is independent of government control, one that concerns itself with the interests of the university itself rather than directly with wider society; with the regulation of truth and reason and with the freedom to speak out publicly without fear of censorship. This role is taken by the faculty of philosophy, which is considered 'lower' than those it regulates by virtue of it falling below, and maintaining a distance from, the exercise of political power.

Kant's faculty of philosophy was further divided into two departments: a department of historical knowledge which includes history, geography and the humanities; and a department of what he called pure rational knowledge, that is, pure mathematics, pure philosophy, metaphysics and ethics. In his earlier work, most notably in *Critique of Pure Reason*, Kant had made a similar distinction between 'scholastic' philosophy, that is, philosophy as a discipline with subject matter that can be taught and learned much like any other discipline; and philosophy as a *conceptus cosmicus*, the overarching, transcendental study of knowledge itself.

Kant is suggesting that we can look at philosophy in two quite distinct ways, perhaps as two separate subjects. On the one hand, philosophy can be considered as the study of 'historical knowledge'; as an academic (or as Kant says, a 'scholastic') subject with a syllabus, a curriculum and a canon of work dating back at least two and a half millennia. It can be divided into 'periods' such as ancient, mediaeval, modern and postmodern; into 'subjects' such as ethics, epistemology and ontology; into 'schools' such as existentialism, phenomenology and logical positivism; into topics such as rationalism and idealism; and these can be conflated to produce programmes of study such as existential ethics or postmodern epistemology. It can also, as I have suggested earlier, be 'applied' to other disciplines such as science, education and, perhaps, nursing. On the other hand, philosophy can be regarded as a form of practical wisdom (*praxis*) rather than a purely academic discipline, as a way of thinking (or, more specifically, as a way of thinking about thinking) rather than a subject to be studied. To do philosophy in this way is to exercise our critical faculties; to turn thought reflexively back on itself. Kant is suggesting, then, that his two departments within the faculty of philosophy represent two different approaches to the subject. The question of what might be the most interesting, important or pressing problems in contemporary philosophy of nursing therefore has at least two different types of answer depending upon where we situate ourselves in Kant's academic structure. It is of course possible to have a foot in each camp, or else to argue that the dichotomy is false, or that the two approaches are aspects of a single indivisible whole. I will, for the time being, continue to treat them as separate, before arguing for the pre-eminence of the latter over the former.

To ask about the most important problems in contemporary philosophy of nursing from the perspective of 'scholastic philosophy' is to ask firstly which nursing problems lend themselves to philosophical analysis (or perhaps, which areas of philosophical inquiry might be applied to the problems of nursing); and secondly, which of those problems, for whichever reason, demand our most immediate and urgent attention. This, to some extent, is a matter of personal interest, opinion, taste and politics. It is also, of course, a matter of economics, of chasing the research grants. On the one hand, not all nursing problems have a philosophical aspect, and not all areas of philosophy are of direct relevance to nursing. June Kikuchi, writing in 1992, suggested that philosophical issues for nursing at the time fell almost exclusively under the remit of either ethics or epistemology, and we might argue that nothing much has changed in the ensuing two decades. So, for example, philosophy of nursing is even today often perceived as not straying far from the study and application of ethics and of making a limited contribution to 'ways of knowing' in nursing. Similarly, the philosophical contribution to nursing research rarely extends beyond a cursory account of deontological ethics and basic epistemology. Kikuchi added that whilst nursing philosophy and nursing ethics are often conflated into a single meaning, she was sceptical about whether nursing ethics even exists as a body of knowledge or an area for study separate from 'ethics proper'; that is to say, whether nursing ethics is anything more than the mechanical application of existing principles and theories to nursing issues. She was similarly sceptical about the epistemology of nursing, but did at least hold out the possibility of, as she put it, raising epistemological nursing questions *with a future*. It seems to me, however, that twenty years on we have made little progress in establishing an epistemology of nursing. Indeed, it could be argued that we are increasingly turning to *medical* research for the gold standard methodologies for producing new knowledge for nursing practice. The fact that we continue to refer to the work of Carper from the 1970s as the touchstone for thinking about nursing knowledge is beginning to suggest that such a quest might be ultimately fruitless, and that perhaps nursing knowledge does not exist as distinct and separate from medical knowledge, however much we might wish it to.

Having dismissed nursing ethics as unremarkable (and perhaps as non-existent) and nursing epistemology as un(der)developed, Kikuchi turned her attention finally to nursing ontology which, she claimed, was conspicuous by its absence from the philosophical study of nursing, but which she considered to be a necessary foundation for all other philosophical questions. Kikuchi applied Aristotle's term 'being *qua* being' (which can also be translated as beings *qua* being and beings *qua* beings) to nursing ontology in terms of nursing *qua* being, which she rather curiously translated as 'nursing as *a* being' and which, she claimed, is the study of the nature, scope and object of nursing. She suggested that our inability to engage fully with questions of nursing ethics and epistemology stem from our failure to engage with nursing *as a being*, that is, to perceive it as a unified, singular whole and to agree on a definition of what it *is*.

My reason for pursuing Kikuchi's line of reasoning is that I believe that she arrived at substantively the correct diagnosis of the state of nursing philosophy, albeit sometimes through arguments that I don't necessarily agree with. That is to say, the most interesting, important and pressing problems in contemporary philosophy of nursing are problems of ontology, although not, in my view, of ontology in Kikuchi's sense of nursing as-a-*being*, nor even of Aristotle's more general study of things-in-the-world. Rather, the most pressing philosophical problems for me are the same ones that troubled Laing more than fifty years earlier; that is to say, the existential problems arising from the study of *our own* being-in-the-world or, to use Laing's term, man in relation to other men.

Laing's starting point was the extreme case of the schizophrenic patient who does not appear to experience himself as a person, and whose distorted self-image is generally reinforced by medical professionals. Laing famously observed that he had great difficulty in observing the textbook signs and symptoms of psychosis when interviewing patients. His first thought was that this was due to a deficiency on his part; that he was not clever or observant enough to spot what other psychiatrists appeared to see. His second thought was that perhaps *they* had got it wrong, before arriving at the realisation that the doctor and the patient form a single existential subject-object Gestalt. That is to say, the 'reality' of the clinical situation is jointly constructed by the doctor and the patient; each perceives the other from their own frame of reference and each responds to the other according to how they imagine that the other is perceiving them (think, perhaps, of Sartre's example in *Being and Nothingness* of the peeping Tom who suddenly switches from observer to observed as he realises that he himself is being watched). The trick is at all times to consider the possibility of the other as a thinking, feeling, autonomous subject and to recognise that the other will often respond to me as nothing more than an object in their own perceptual field.

Laing's project, which he articulated in the terms and language of existential phenomenology, was to move beyond the traditional 'objective' doctor-patient relationship towards a subjective hermeneutic understanding from the patient's point of view. Whilst a *full* hermeneutic understanding is rarely if ever possible, Laing argued that we must nevertheless resist the temptation to assess, judge and respond to the other according to our own medico-bio-scientific categories. The least we can do is to reach out and attempt a person-to-person existential grasp of the lived experience of the other in the form of a therapeutic relationship. Whilst there are psychological and sociological aspects to this undertaking, it is at heart a philosophical venture, and it seems likely that Laing was influenced here by Sartre's notion of the transcendent ego; that is, the argument that the 'I' or ego is, in his own words, situated 'outside *in the world*; it is a being in the world, like the ego of another'. This supposition to some extent circumnavigates the 'other minds' problem and opens up the possibility of authentic and largely unmediated contact between myself and others 'out there' in the world. We might say, then, that the most fundamental and pressing philosophical problem in the field of nursing is the question of otherness; of attempting to understand and engage with our patients and colleagues. Whilst Laing explored this problem specifically within the relationship between the psychotic patient and the psychiatrist, I would suggest that it has a wider relevance that extends to the nurse-patient relationship in general (that is, not only in psychiatric nursing), indeed, that it is *fundamental* to the practice of nursing.

The importance of an existential hermeneutic understanding of the other was, for me, brought into sharp relief by the recent and very shocking case in the UK of a 22 year-old man who died from dehydration whilst being treated in hospital. This example can be added to a growing list of cases of neglect and failure to care, mostly by nurses but sometimes also by other health care professionals. Increasingly, the response from the general public and from patient groups is that nurse education has become too academic and nurses too highly qualified, resulting in a lack of common sense and basic compassionate care at the bedside. Sartre's work, however, suggests the opposite. His position is that the 'natural attitude' (the common-sense view) is of other minds, other Egos, as largely inaccessible to our own consciousness and hence unknowable (and even, perhaps, unverifiable). We tend most of the time to view others merely as objects in our perceptual field, and when we do view them as autonomous subjects, it is rarely from a position of empathic understanding. As Laing pointed out, our failure to understand the life-worlds of others is largely due to our experience of them as fundamentally separate and removed from our own experiences. In the example cited earlier, the agitated and aggressive behaviour exhibited by the dehydrated patient was wrongly interpreted because no attempt was made to understand the world from his perspective. The remedy for this gross lack of imagination is not less education but more. Compassion cannot be taught, but the knowledge and understanding necessary for its recognition and expression most certainly can. The application of common sense is not the solution but part of the problem.

For me, the practice of nursing is fundamentally concerned with the development of caring human relationships, and this requires a deep philosophical understanding of the nature of our own being *qua* being and that of our patients. All other problems in contemporary philosophy of nursing, including matters of ethical conduct and questions about the nature of nursing knowledge, are secondary to this fundamental ontological concern and, I would suggest, can only fully be addressed within the context of what Laing referred to as the existential-phenomenological science of persons. Put another way, we must first establish the principles of an authentic and respectful patient-nurse relationship before we can even begin to formulate questions about what it is possible and desirable to know and how we should behave in the context of that relationship.

3. What, if any, practical and/or socio-political obligations follow from studying nursing from a philosophical perspective?

From Kant's perspective of 'scholastic philosophy', that is, philosophy as an academic discipline of the same order as other disciplines in the university, I have suggested that contemporary philosophy of nursing must begin with ontology, that is, with questions of being *qua* being, which is to say, with questions of human relationships. The practical, socio-political and ethical obligations that follow from my suggested focus for philosophy of nursing are largely self-evident and have been outlined briefly above. I suggested also that we have barely started; that philosophy of nursing in a disciplinary sense is a very long way from being fully realised. However, any attempts to establish philosophy of nursing in a disciplinary sense must, like all aspects of academic nursing, proceed from an ethico-ontological desire to make a difference to nursing practice rather than from a political move to stake out a territory from which to impose raw theory on a neophyte discipline.

We have seen that Kant also proposes philosophy as a *conceptus cosmicus*, that is, as a transcendent discipline, and it is perhaps here that philosophy can make the most immediate contribution to the study and practice of nursing. To consider philosophy as a *conceptus cosmicus* is to consider how philosophers think rather than what they think about;

that is to say, it is to consider philosophy as a practice rather than as a theoretical subject to be studied. It is, of course, useful and relevant to study philosophy from a theoretical perspective, to analyse key ideas and to compare schools of thought, just as it is useful and relevant to study music or fine art in the same way. However, we might argue that philosophy in its 'pure' sense is a practice more akin to art and music than to geography and sociology. Whilst it is possible to study art, music and philosophy 'scholastically' as academic subjects, nevertheless studying the history or the theory of art is very different from practising as an artist. And just as we might be reluctant to refer to a university lecturer with a degree in the history of art or the theory of music as an artist or a musician, so I would suggest that the philosophy lecturer is not *de facto* a philosopher. Conversely, however, the geography lecturer might well be a philosopher in Kant's 'pure' sense, even if they are not teaching any academic content associated with the discipline of philosophy.

To philosophise in this sense is simply to think, in Ryle's sense of *just thinking*. Ryle distinguished between thinking as a component of another activity (he gives the example of playing tennis) and thinking as an activity in its own right (the example of Rodin's *Le Penseur*). In the context of my earlier example, the geography lecturer and the philosophy lecturer will each think 'scholastically' about their academic disciplines, but they might or might not also simply *think* (or, we might say, reflect or philosophise). In *The Differend*, Lyotard made a similar distinction in this respect between the intellectual and the philosopher. Whereas the intellectual has a disciplinary agenda (in both senses of the term) that includes closing down discussion and debate by imposing established (and establishment) solutions, the philosopher is concerned only with keeping debate alive, that is, *to save the honour of thinking*. Badiou similarly distinguishes between the 'TV philosopher' who talks about society's problems, and the 'genuine philosopher' whose job is not to *solve* problems but to *pose* them. The genuine philosopher does not have a home discipline, whether a department of philosophy or a department of nursing (or, rather, their home discipline is more or less irrelevant to their work as a philosopher). To quote Badiou (2009), 'the philosopher intervenes when in the situation – whether historical, political, artistic, amorous, scientific... – there are things that appear to him as signs, signs that it is necessary to invent a new problem'. This is philosophy not as a subject, not as a discipline, but as a duty that we *all* owe to the academy and to ourselves.

4. In what ways does your work seek to contribute to philosophy of nursing?

I do not teach philosophy as a 'scholastic' subject beyond the occasional lecture on poststructural or postmodern approaches to nursing. However, I have published a number of journal papers and several books, including *Research, Truth and Authority* (2000), *Deconstructing Evidence-Based Practice* (2004) and *The University in Dissent* (2013), which have applied philosophical ideas and thinking to research, practice and education respectively. However, any contribution that I might make to philosophy *of* nursing (as opposed to philosophy *for* nursing) will be in the wider sense of reinstating Badiou's notion of *reckless* philosophy (that is, a philosophy that is concerned with means rather than with ends) back into a university that long ago closed down its philosophy department. That is to say, I attempt to practise philosophy in *how* rather that in *what* I write and teach; I attempt wherever possible to *ask* rather than to *answer* questions and to *create* rather than to *resolve* problems. In all of this, I try in everything I do to follow Jacques Derrida's dictum of striving at all times to keep open and alive questions of importance and relevance to nursing and to the wider academy, and of never taking 'no' for an answer: 'For there must not be a last word – that's what I'd like to say finally'.

5. Where do you see the field of philosophy of nursing to be headed, including the prospects for progress regarding the issues you take to be most important?

To paraphrase Ghandi when asked what he thought of Western civilization, I think philosophy of nursing would be a good idea. However, as I suggested earlier, we remain some distance from a bona fide philosophy *of* nursing in a disciplinary sense as opposed to philosophy *for* nursing, that is, the imposition of ideas from one discipline on another. I believe that a philosophy of nursing in a scholastic or disciplinary sense is achievable and that we are making progress towards it: we have a journal, an annual conference and a number of nascent programmes of research which are making great strides in fully integrating philosophical ideas and concepts into mainstream nursing concerns. That is to say, we are beginning to look at nursing *philosophically*. So long as a philosophy of nursing proceeds from, and is grounded in, a commitment to understand others and to make a difference 'out there' beyond the academy, then wherever it finds itself will be the right place.

References

Badiou, A. 2009. "Thinking the Event." In *Philosophy in the Present* edited by P. Engelmann. London: Polity Press.

Badiou, A. 2011. *Second Manifesto for Philosophy*. London: Polity Press.

Derrida, J. 1992. *Afterwords*. Tampere: Outside Books.

Freshwater, D. and G. Rolfe. 2004. *Deconstructing Evidence-Based Practice*. Abingdon: Routledge.

Kant, I. 1929. *Critique of Pure Reason*. London: Macmillan.

Kant, I. 1992. *The Conflict of the Faculties*. Lincoln, Nebraska: University of Nebraska Press.

Kikuchi, J. 1992. "Nursing questions that science cannot answer." In *Philosophic Inquiry in Nursing* edited by J. Kikuchi and H. Simmons, 26-37. Newbury Park California: Sage.

Laing, R. D. 1969. *The Divided Self*. Harmondsworth: Penguin Books.

Lyotard, J-F 1988. *The Differend: Phrases in Dispute*. Minneapolis: University of Minnesota Press.

Rolfe, G.

2000. *Research, Truth and Authority: Postmodern Perspectives on Nursing*. Basingstoke: Macmillan.

2001."Postmodernism for Healthcare Workers in Thirteen Easy Steps." *Nurse Education Today* 21: 38-47.

2004."Deconstruction in a Nutshell." *Nursing Philosophy* 5: 274-6

2009."Some Further Questions on the Nature of Caring." *International Journal of Nursing Studies* 46: 143-6.

2012."Cardinal John Henry Newman and 'The Ideal State and Purpose of a University." *Nursing Inquiry* 19: 98-106.

2013."Thinking as a Subversive Activity: Doing Philosophy in the Corporate University." *Nursing Philosophy* 14.

2013. *The University in Dissent: Scholarship in the Corporate University*. Abingdon: Routledge (Forthcoming).

Rolfe, G. and L. Gardner. 2006. "'Do not ask who I am...' Confession, Emancipation and (Self)management Through Reflection." *Journal of Nursing Management* 14: 593-600.

Rolfe, G. and L. Gardner. 2006. "Towards a Geology of Evidence-based Practice." *International Journal of Nursing Studies* 43: 903-13.

Rolfe, G. and L. Gardner. 2006. "Education, Philosophy and Academic Practice: Nursing Studies in the Posthistorical University." *Nurse Education Today* 26: 634-9.

Ryle, G. 1971. "Thinking and Reflecting." In G. Ryle *Collected Essays Volume II*, 465-479. London: Hutchinson.

Sartre, J-P. 2004. *The Transcendence of the Ego*. Abingdon: Routledge.

19

Trudy Rudge

Professor of Nursing (Chair, Social Sciences and Humanities)

University of Sydney, Sydney, Australia

Beware that you do not lose the substance by grasping at the shadow (Aesop)

1. How were you initially drawn to philosophical issues regarding nursing?

Bodies, motives (emotions, ethics, aesthetics), spaces and existences are some matters that sent me to philosophy to provide accounts, explanations and further thoughts on nursing. In my thinking about bodies, I was first drawn to continental philosophers such as Merleau Ponty, who has a considerable *oeuvre* that addresses the mind/body problem. Primarily, I am more drawn to his last and incomplete work on *The Visible and Invisible* with his formulation of flesh as an open and intersubjectively experienced body (Merleau-Ponty 1968). But as my major exploration of this was post-1980s, Foucault (1977, 1980, 1986, 2001) and others such as Elizabeth Grosz (1994) and Moira Gatens (1994) as feminist philosophers using his work, have had a major impact on my thinking about bodies. Foucault's work on the panoptic disciplinary society (1977) and the more nuanced account of governmentality (1986, 2001) are an on-going presence in how I think about bodies (both of nurses and patients). Foucault and bodies and their care is threaded through my work on burns care, mental health care and nursing as well as a concern for finding an ethical position from which to work as an academic.

In thinking about skin and bodies, the issue was to explore how to think about what Francine Wynn (2002) talks about as witnessing. This is counter to much of the thinking about nurses' activities that explore what nurses do with patients. Such medicalised discourses act to exclude the entry of the uncertain and partial into any calculus of treatment. This also leads to a focus on a singular way of interpreting the human body and nurses' interactions with it (Shildrick 2005). Instead, using philosophical works has allowed a focus on how and in what way nurses and patients interact because they are embodied.

The hope from such thinking is to expose how binary structures such as mind/body, inside/outside, abnormal/normal are rendered porous in the healthcare situation, defacing nursing care practices. Following this, I am interested in care as an ethical set of practices rather than as a set of dogma or worse, as a vocation flowing automatically from our professional standing with its rules and codes of conduct. Rather, I believe that care requires recuperation where explorations of it are based in everyday care and micro-ethical concerns. As Margrit Shildrick (2005, 17) puts this, such a recuperation would be based on 'post-modernist bioethics [demanding]... an openness to the risk of the unknown, a commitment to self-reflection, and a willingness to be unsettled [in an] enterprise of high responsibility'. More importantly, care is not merely located as affective but is about the caring practices that are embodied and knowledgeable – as Purkis and Bjornsdottir (2006) claim, knowledge in action.

In seeking to find a place to speak about nurses as knowledge workers and practitioners I have had to move to Foucault's later work, Deleuze (1997) with Félix Guattari (2004), and more lately to Slavoj Žižek to think about the impact of the violence of the neo-rationalist health care policies and practices on nurses' work in health care in the variety of settings where nurses now work. To explore motivation or intentions of nurses it was important to use the concept of desire following Spinoza (2000), Žižek (2009, 2012) and Deleuze (1997) to develop an ethics of the body inflamed by desire, with ideas of being connected rather than separated from each other, based on an exploration of actions rather than principles – moved by continental philosophy that provides manifold points of view about nursing and nurses' activities. A focus on the rules of ethics has not proven to be effective for nurses to be able to undertake ethical care that is spoken from a knowledgeable position – a position where the nurse is responsible for their own behaviour and position first. Motivations for caring therefore are to be thought through from the understanding that nurses are responsible for caring and caring practices. Motivations for taking such position are productive rather than merely a mechanical response to rules or to actions that may endanger one's registration (Bauman 2008a). Ethical responses through the lens of mere regulatory statements fail to respond to the micro-realities of nursing practice or indeed be sufficient for a nurse to take action (Nelson and Purkis 2004).

2. What, in your view, are the most interesting, important, or pressing problems in contemporary philosophy of nursing?

In her notes on nursing, Nightingale (1992, 6) states that while it is believed,

> that every woman makes a good nurse. I believe on the contrary, that the very elements of nursing are all but unknown. By this I do not mean that the nurse is always to blame. Bad sanitary, bad architectural, and bad administrative arrangements often make it impossible to nurse. But the art of nursing ought to include such arrangements as alone make what I understand by nursing possible.

It is galling to note that most of this quote from Nightingale would not be considered out of place in contemporary discussions concerning nursing and its place in health care. It is clear that the most pressing problems for nursing are both interesting and important, and of long standing. They remain located in this on-going discussion about what counts as nursing in the health care systems around the world – what it is and what it is not. Is it sufficient for nurses to be recognised as the glue of the system (Sandelowski 2002) or the organisers of the system (Rankin 2009)? If we take away the work that nurses do for 'the system', what would be left?

I have some sympathy with those who see that what we need is more nurse-led research into the care they provide for people who come to their area for care. We do need to think about what nurses do in their clinical practice more clearly and to validate nursing effort as part of making 'arrangements' to artfully nurse. There are many rewards for those thinkers who do this, while we need to be clear about what counts as measurement as the debate on what counts as evidence has shown (Holmes *et al.* 2006). Hence, it is central in accounting for nursing effort to be mindful of reductionist and simple responses to the complex problems that largely make up most of nursing's interests in terms of research. And so the interesting problem is how to bridge that classical divide about what is clinical and what is theoretical. How can nurses who have the time to think about nursing contribute to the support of nurses who increasingly have little time to think, or whose thinking is stopped or undervalued; and provide meaningful witness to the work of nurses.

The issue is not to get the nursing voice heard, but rather for nurses' work in the system to be acknowledged as crucial to the care of the people who come before us in the health care system. Unfortunately, I am aware of only pockets of activity that focus our thinking or philosophising on nurses and their thinking about themselves before we learn to provide or ride into rescue our clientele. Nurses have not taken on board the ideas from Foucault about care of the self (1986, 2001), but have rather rushed to caring for others, and still do. As I think about the position of nurses, as professionals in the health care system, I notice that the move to 'professional' is entangled now with regulation in many jurisdictions, as if this is the key mechanism for being professional – policing ourselves. Such an entanglement is problematic when there are many aspects of nurses' work that are affected by the organizations where they work rather than being derived from nurses' efforts at defining their places in such organizations. Therefore one of the most pressing problems for nurses, who in large part remain professionals who work in large health care agencies, is how to set their agendas when most of such organizations are driven by bureaucrats with a single agenda – to reduce spending, rather than develop their workforce[1] (Rudge 2011).

A role for faculty working in academic nursing is to form alliances with those who work in these straitened locations, and to develop ways of speaking out for the pressures that militate against nurses undertaking their work as they believe they should. The problem is to find ways to participate in this without going down the pathway as mere writers of policy (a political *cul de sac* that Foucault (2001) warns us about in *Fearless Speech*). Writing policy is writing for government, not for nurses and is meant to be part of what it is to govern, which may be why such writing and positioning is seductive. Moreover, many of the guidelines, protocols, codes and regulation seek to control nurses and nursing without providing opportunities to extend nursing knowledge or practices. The complexity of new ways of organising society, with its consumer oriented mentality, instead require a lightness of step not seen in the fixedness of nursing in contemporary health care organizations (Bauman 2008b) as represented by regulatory agenda.

---

[1] In an interview of Professor John Buchanan with Australian Broadcasting Corporation News 24 Breakfast, Buchanan discussed the characteristics of the Australian workforce, and he pointed out that while the focus is on production in terms of manufacturing and the small percentage of the workforce in mining, that agriculture, health and education make up the major proportions of the workforce. Health makes up 15% of the workforce in Australia, hence a major employer – three times that of mining. This situation would not be uncommon in other jurisdictions.

3. What, if any, practical and/or socio-political obligations follow from studying nursing from a philosophical perspective?

I have an aim and a place or two to stand. I want nurses to think about what they do and how they go about it. I believe that philosophical perspectives provide the tools with which to work, to have a collection of places that we can stand that first positions us to listen. For philosophy to assist with this, the obligations that follow from thinking about nursing from a philosophical perspective are to provide explanations about the situation for nursing at this moment, and into the future – this is a view that is much reduced at the moment.

As Freidson (2001) warns in his prognostications about the future of professions, he is concerned about how many professions are being influenced by narrowing of their range of activities, broadening of the range of their remuneration, and dilution by the use of paraprofessionals or lower status workers to undertake some of their tasks. Moreover, in the drive to meet the needs of governments with professionals' work linking strongly to the needs of industries (or perhaps in the latest parlance, stakeholders) may not benefit the future. He states:

> Emphasis on service to the practical needs of the state, the discovery and development of profitable goods and services by private capital, and satisfaction of what the public believes to be its needs may restrict and narrow the direction of the development of knowledge by limiting it to what is presently known and believed and what can be anticipated by the passive projection of trends. No one can predict what may be lost when idle curiosity and purely theoretical interest are discouraged, but it could be substantial. (Freidson 2001, 213)

As researchers, thinkers and practitioners of nursing it is central to our endeavours that we find a way to do this that provides nurses researching nursing with a meaningful framework to understand and 'witness' nurses' work. Moreover, it seems important that the discussion be understandable, not always easy to read but make sense for some people. In talking about nursing I am constantly trying to make links between the clinic and the academy, so that nurses who work every day in these areas have our understanding. This has a political intent. What is happening in the clinic and the academy to nurses has to be viewed as one – not the same but as a clinically derived academic and thinking problem. In such a location, strategies require to be developed for what is being said, an account provided for those who are silenced and also accounts as to how this silence works against the people that come to or in front of nurses in the work they do. As an academic I have the opportunity to speak about issues as I see them affecting nurses wherever they work – the protection of academic freedom allows me to write and research those things which can support nurses' work or uncover nurses' concerns. As I have noted at the beginning of the chapter, the politics of developing philosophies of nursing requires a position from where I explore what causes me some discomfort, not just about what I see or hear, but how I am to account for it. My position affords me a way of extricating myself enough to obtain a view that I can speak about, but not always with ease (Foucault 2001).

4. In what ways does your work seek to contribute to philosophy of nursing?

I would not think that what I do is seek to contribute to a philosophy of nursing, but rather a set of papers and writings that seek to bring philosophy to exploring nursing and the complex situation that all nurses find themselves in. Like Aesop suggests above, I like to explore the shadows for the substance that may be lurking there, hidden by our focus on the shadow. This means that what I like to focus on is some matters that have come to be seen as central to nursing so as to be made visible, or to be heard or to prove that nurses are responsible citizens. I suppose the philosophy I want to contribute to is one that looks behind or sideways (Žižek 2009), to expose what is enrolling nurses to enact policies and practices that have the potential to be both good and bad for 'nursing'. The point that has to be made is to expose how we get seduced by power or its potential. Using philosophy allows ways to think about nurses and nursing in a variety of settings, in terms of relations between nurses and then between them and their clientele.

Each country has its issues about education, workplace and ethico-legal concerns. The development of philosophy allows nurses to question solutions held out to them by nursing leaders or academics. We have had many solutions offered to nursing but often the implications of their use are not always so easily seen except in hindsight. The central element of this questioning is to explore what the effects of taking up a particular position means for nursing practices and nursing's positioning in the health care system. Using philosophical concepts and approaches underpins much of the work to explore nursing, particularly when concepts are not taken up critically and we fail to explore the implications for nursing more generally. There are several concepts that continue to be taken-for-granted as essential to nursing yet have led us back to doing the work of the system rather than for nursing. Instances such as the use of competencies for regulation; evidence based practice and the evaluative strategy of randomised controlled trials of complex nursing interventions; practice development for the development of nursing practice; and the use of lean principles for organising nursing are all concepts requiring vigilance.

When we take up such concepts, it is important to question how such concepts or tools advance nursing and nurses' place in the provision of clinical care in the health care system. Also it remains unclear how or in what way administrative and research thinking contributes to nursing knowledge and value. Clinical knowledge is important, yet it remains difficult to see how some is from a nursing perspective. Moreover, many clinical tools have political effects on both nurses and patients. The political aspects of tool development are seldom addressed. Moreover, a philosophical approach provides a framework for understanding what is happening in nurses' work where economic and political agenda play out across their workplaces. If we fail to see that clinical nursing happens in a workplace affected by matters that are not clinical, then analyses of nurses where they practice will fail to expose the full dimensions of nursing work. In particular, it is important to recognise the changes being wrought in the world of work under advanced capitalism as it breaks collectives and inflexible workplaces, where the consequences for those with a more fixed work location rather than possessing the fluidity of the flexible worker, are yet to be worked through (Boltanski and Chiapello 2007).

5. Where do you see the field of philosophy of nursing to be headed, including the prospects for progress regarding the issues you take to be most important?

The directions of philosophy of nursing are exciting. Certainly the writing that is seen in papers using philosophy and directing nurses' thinking about their practices or the education of nurses are covering a wider range of thought on the characteristics of nursing and its values. It is possible that we are now thinking more about what we are doing and questioning whether what is offered as a direction serves our agenda or is more about the health care system or clinical outcomes rather than contributing to the development of nursing as a discipline and practice. More than any other frameworks perhaps the humanities disciplines are able to analyse nursing outside of specialist or practice settings within the discipline. Such opportunities are less available, where more programmatic approaches are becoming more common. Philosophy of nursing is therefore required, where it is understood that

> critique does not consist in saying that things aren't good the way they are. It consists in seeing on what type of assumptions, of familiar notions, of established, unexamined ways of thinking the accepted practices are based. (Foucault 2000, 456)

To actively promote a philosophy of nursing that explores the assumptions of concepts or ideas that are used in nursing seeks to provide nurses with a location from which to develop nursing knowledge and work.

Philosophy of nursing (perhaps like other humanities frameworks such as history of nursing) still affords nursing the best view across the various settings and locations where nursing is increasingly fixed and less aware of what is happening across or outside nursing. Philosophy therefore can obtain more purchase when it can promote how this is important to nurses in a variety of locations with the proviso that such analyses always focus on contextual issues to see how difference comes about. As nurses in all settings work in surroundings of high responsibility, a philosophy of nursing increasingly aims at providing an ethical stance to work as a nurse, and to resist the many organisational schemes that subvert nursing values. More nurses are aware that their values are not always shared in the organisations that provide health care; more philosophers are providing ways for nurses to act against such situations, whether collectively or on their own. Further, such writing provides insights into the contradictions that abound at this moment of great change in the world of work, with its destabilisation of collective approaches to bargaining, the growth of individualization and fluidity occurring just as nurses are called to follow fixed guidelines, protocols and rules that contradict such changes in the models of work and care (Bauman 2008b). Increasingly, the philosophy of nursing explores these troublesome and challenging aspects as well as the opportunities that such openings provide.

References

Bauman, Zygmunt. 2008a. *Does Ethics Have a Chance in a World of Consumers?* Cambridge, MASS: Harvard University Press.

Bauman, Zygmunt. 2008b. *The Art of Life*. Cambridge: Polity.

Boltanski, Luc and Eve Chiapello. 2007. *The New Spirit of Capitalism*, translated by Gregory Elliott. London: Verso.

Deleuze, Gilles. 1997. *Essays, Critical and Clinical*, translated by Daniel Smith & Michael Greco. Minneapolis: University of Minneapolis Press.

Deleuze, Gilles and Félix Guattari. 2004. *Anti-Oedipus: Capitalism and*

Schizophrenia, translated by Robert Hurley, Mark Seem, and Helen R. Lane, preface by Michel Foucault. London: Continuum.

Foucault, Michel. 1977. *Discipline and Punish: the Birth of the Prison,* translated by Alan Sheridan. London: Tavistock.

Foucault, Michel. 1980. *Power/Knowledge: Selected Interviews and Other Writings 1972-1977,* edited by Colin Gordon. New York: Pantheon.

Foucault, Michel. 1986. *Care of the Self*, translated by Robert Hurley. New York: Pantheon.

Foucault, Michel. 2000. "So Is It Important to Think?" In *Power: Essential Works of Foucault 1954-1984, Volume 3,* edited by James D. Faubion, Translated by Robert Hurley and others. New York: The New Press.

Foucault, Michel. 2001. *Fearless Speech,* edited by Joseph Pearson. Los Angeles, CA: Semiotext(e).

Friedson Eliot. 2001. *Professionalism: the Third Logic*. Chicago: University of Chicago Press.

Gatens, Moira. 1994. *Imaginary Bodies: Ethics, Power and Corporeality*. London: Routledge.

Grosz, Elizabeth. 1994. *The Volatile Body: Towards a Corporeal Feminism*. Sydney: Allen & Unwin.

Holmes, Dave, Stuart J. Murray, Amélie Perron, and Geneviève Rail. 2006. "Deconstructing the Evidence-based Discourse in Health Sciences: Truth, Power, and Fascism." *International Journal of Evidence Based Health Care* 4: 180-186.

Merleau-Ponty, Maurice. 1968, *The Visible and the Invisible,* edited by Claude Lefort, translated by Alphonso Lingis. Evanston, IL: Northwestern University Press.

Nelson, Sioban and Mary Ellen Purkis. 2004. "Mandatory Reflection: the Canadian Reconstitution of the Competent Nurse." *Nursing Inquiry* 11: 247-257.

Nightingale, Florence. 1992. *Notes on Nursing: What It Is and What It Is Not,* The Commemorative Edition. Philadelphia: Lippincott.

Purkis, Mary Ellen and Kristin Bjornsdottir. 2006. "Intelligent Nursing: Accounting for Knowledge as Action *in* Practice. *Nursing Philosophy* 7: 247-256.

Rankin, Janet. 2009. "The Nurse Project: An Analysis for Nurses to Take Back our Work." *Nursing Inquiry* 16: 275-286.

Rudge, Trudy. 2011. "The Well-run System and Its Antimonies." *Nursing Philosophy* 12: 167-176.

Shildrick, Margrit. 2005. "Beyond the Body of Bioethics: Challenging the Conventions." In *Ethics of the Body: Postconventional Challenges*, edited by Margrit Sheldrick and Roxanne Mykitiuk, 1-26. Cambridge MASS: Massachusetts Institute of Technology Press.

Sandelowski, Margarete. 2000. *Devices and Desires: Gender, Technology, and American Nursing*. Chapel Hill: University of North Carolina Press.

Spinoza, Benedict. 2000. *Ethics*, edited and translated by George H. R. Parkinson. Oxford: Oxford University Press.

Wynn, Francine. 2002. "Nursing and the Concept of Life: Towards an Ethics of Testimony." *Nursing Philosophy* 3: 120-132.

Žižek, Slavoj. 2009. *Violence: Six Sideways Reflections*. London: Profile Books.

Žižek, Slavoj. 2012. *Organs Without Bodies: On Deleuze and Consequences*. London: Routledge Classics.

20

Anneli Sarvimäki

The Age Institute

Asemapäällikönkatu 7, Helsinki, Finland

---

1. How were you initially drawn to philosophical issues regarding nursing?

My first interest was philosophy, not nursing. My interest in philosophy goes back to high school, where I came to read a textbook introducing philosophy and a book presenting Sartre's philosophy. I have carried with me the existentialist themes ever since. While still in high school I also published my first article in the daily press; it concerned different levels of experiencing music. When I went on to university I chose philosophy as one of my study subjects. That gave me an opportunity to dig deeper into existentialism and all the other philosophical subareas. So, I had already completed my MA degree with studies in educational sciences, philosophy and psychology when I became a nursing student. The reason why I took on nursing was that I had been recruited to teach philosophy, ethics and pedagogy at a nursing school while I was still studying for my MA degree. I became interested in nursing as a discipline with great philosophical potentiality. But I realized that in order to be able to contribute philosophically to nursing, I had to get thorough knowledge of nursing, both theoretically and in practice. And so I became a nurse and specialized in psychiatric nursing. I wrote my thesis together with another student, Bettina Stenbock-Hult, with whom I have been co-writing ever since. Our thesis was about intuition in nursing, and later on we expanded the topic into an article (Sarvimäki & Stenbock-Hult 1996).

Thus, I initially entered nursing from a philosophical standpoint and was trained in philosophy when I started my nursing studies. I published my first texts (in Swedish and Finnish) on philosophical issues regarding nursing in the late 1970s and early 1980s. They concerned ethics and conceptions of man (philosophical anthropology). I also wrote a teaching compendium outlining philosophical areas relevant for nursing in the late 1970s, and in 1985 I published the first textbook on nursing ethics written in Finnish together with professor Hertta

Kalkas (Kalkas & Sarvimäki 1985). The book was also translated into Swedish. Having completed my nursing education I continued with my academic studies. For my Licentiate and PhD degrees in the 1980s, I chose to study topics relevant to nursing from a philosophical point of view: health, ethics, creative interaction, practical knowledge, intuition (Sarvimäki 1988a, 1988b).

2. What, in your view, are the most interesting, important, or pressing problems in contemporary philosophy of nursing?

I cannot say that one philosophical problem is more important or pressing than another. Every aspect of nursing can expand into as interesting philosophical problem. What I can see, however, when I view the state of the art in the philosophy of nursing, is that some areas have been more investigated while others are still waiting to be studied. Nursing has traditionally been conceived as a practice discipline, where the individual patient, the relationship between nurse and patient, and ethical values have been predominant features. This has influenced what kind of questions nursing philosophers have dealt with. Ethics is probably the most studied sub-area of philosophy of nursing, and it was also my entrance-gate to nursing and philosophy of nursing. The emphasis on ethics is understandable, because ethics has always been one of the cornerstones of nursing, so it is natural that it should also attract philosophical interest. Another popular topic has been epistemology, which attracted my interest too when I got a chance to probe deeper into philosophy as a discipline. One explanation for the growth of interest in epistemological questions, at least in the Scandinavian countries, might be that nursing was developing into an academic discipline at the same time when there was much interest in practical knowledge, tacit knowledge, experiential knowledge and knowledge-in-action. All these forms of knowledge were relevant to nursing, being a practice discipline. Furthermore, the discussion on the epistemological base of qualitative and quantitative research traditions also influenced nursing, causing a debate between research paradigms and the nature of nursing science and its epistemological basis. Thus, philosophy of nursing is not an isolated discipline. It has been, and still is, very much influenced by what is going on in other disciplines.

In order to say something about important problems in contemporary philosophy of nursing, I will start by outlining on a more general level the tasks of philosophy in relation to nursing, since what problems you see as the most pressing right now depends partly on your view of philosophy.

Philosophy can be said to have three main tasks in relation to nursing (Sarvimäki 1999). The first task is problematizing, asking questions.

This means pointing out that values and ways of thinking that we might have taken for granted are actually quite problematic. For instance, one commonly adopted view in nursing research seems to be that the choice of research methods has to be based on an epistemological standpoint, which, in turn, must be based on an ontological assumption of the nature of nursing. I consider this a very problematic view. If research methods are rooted in ontology, can I then use one method in one study and another method in another study without changing my ontological assumptions? And is it reasonable to assume that researchers should have to, or even be able to, change ontological views from one study to another? Another problematic view I have encountered is the view that man is or consists of body, soul and spirit, and that all human beings are religious even though they are not aware of this themselves, and that everyone believes in something. This view has so many problems that it is difficult to know where to start. Firstly, the expression *is* or *consists of* indicate an essentialist view of Man with all the problems inherent in that view. The claim that all human beings are religious seems to be a proposition needing a process of infinite verification, and the concept of belief in this view needs specification. Does *believe* have the same meaning in the sentences *I believe it will rain tomorrow* and *I believe in a higher spiritual power?* I am not going to go deeper into these problems in this context, I just wanted to show an example of the problematizing task of philosophy. The point is to show that certain values or conceptions lack clarity, are illogical and tend to have untenable consequences.

The second task of philosophy is a kind of digging, revealing and "cleaning work" (cleaning up messy thinking) that goes further than problematization. It involves unraveling the values, assumptions and thought structures that underpin our ways of talking, writing and doing nursing, criticizing unethical or illogical forms and offering better thinking. I might, for instance, claim that inherent in the values of autonomy and individualism, commonly cherished in nursing, risk reducing both patient and nurse. They do not reflect the complex social relationships in which people live their lives. They either have to be re-defined or complemented by other values. Another example of this task is to show how the logic of the industrial process and marketing world has corrupted nursing and also health and social care in general. At least in the care of older persons in Finland, the planning and decision making rely more and more heavily on the exact measurement of the client's functional capacity, the definition of specific care tasks, measurement of time used for every task, and evaluation of efficiency. This is often combined with the logic of the market where old persons are reduced to pawns in the process of buying, selling, outsourcing and competing. But what are the consequences of this logic for the older persons, their relatives and the nurses? The production-marketing logic does not catch the everyday life and experiences of the older person, his or her fears and hopes, struggling and coping. It does not catch the vulnerabilities, hopes and fears of the nurse encountering the old person. Shouldn't the goal of care be conceived in terms of a life as good as possible for the older person? In what way do the measurements and definitions of specific tasks contribute to this? Shouldn't the process of care rather be conceived in terms of creating opportunities for meaningful encounters, positive experiences and personal growth? One could argue, of course, that the production-marketing logic applies to the administrative and economic level, while the logic of interactive action, experiences and meaning applies to the world of nurse-patient encounters. But is it possible to apply the logic of creative interaction within a process that is administrated by the logic of production and marketing?

The third task involves outlining and arguing for reasonable and ethically legitimate ways to conceive, carry out and study nursing. This endeavour can target nursing as a process and phenomenon as a whole, in which case we are talking about outlining *a* philosophy of nursing. Philosophy becomes both the process and its result. We have examples of this from the 1970s on. The philosophical conceptions can also restrict themselves to a sub-area of nursing, e. g., the body, encountering death, patient self-determination. The difference between the encompassing philosophies and the restricted philosophical conceptions or theories can be paralleled with the difference between grand theories and middle-range theories in theory construction. The crucial point, however, is that the philosophy as a product – grand or restricted – has to be based on philosophically sound argumentation, otherwise it is just a personal belief system or a set of ideas randomly put together.

I think that a reasonable amount of philosophizing has been done with respect to the third task. Less has been done on problematizing, digging and revealing. Thus, I cannot pick out a set of specific pressing problems in nursing philosophy. I would rather call for a more varied way of doing nursing philosophy, more problematization, digging, revealing and presentation of grounded and logical argumentation.

3. What, if any, practical and/or socio-political obligations follow from studying nursing from a philosophical perspective?

One obligation that follows from all philosophy is the obligation to clarify our own thoughts and to do better thinking. And, hopefully, clearer thinking leads to better grounded action. But the question is, of course, larger than this, and it can be related to the three tasks of philosophy that I outlined above. Problematizing, asking questions, will probably not as such lead to any practical or socio-political obligations. Usually, problematizing leads to confusion. It can show us that our ways of thinking in and about nursing and our ways of conceiving nursing are not as self-evident as we might think. We start to question our thoughts, ideas and values and may feel the ground shaking under our feet. But this does not necessarily lead to any practical or socio-political obligations. It only means that we have to think more and think deeper. Confusion is a good state of mind. It pushes us towards greater clarity.

The second task of philosophy may well have practical consequences. If our revealing and cleaning work shows that the values and logic that we have based our nursing on are untenable or corrupt, then we will have to start thinking in terms of alternative values and logic. A serious questioning of the present production-marketing logic in health care, which I outlined in my answer to the previous question, is bound to have socio-political obligations as I am sure many other philosophical endeavours have. The conceptualization of nursing and health care in terms of communicative action, creative interaction and shared responsibility on a societal level might give us alternative ways of thinking, planning and acting in nursing.

4. In what way does your work seek to contribute to philosophy of nursing?

Contributions to philosophy of nursing cover in my view more than just conducting philosophical studies and have them published. Promoting philosophical awareness among nurses and nurse teachers as well as implementing philosophical concepts in research and practice are also ways of contributing. I can see three lines along which I have tried to contribute to philosophy of nursing: teaching, philosophizing, and conducting empirical research based on philosophical themes.

The first line is connected to my profession as a teacher in colleges and universities in Finland and Sweden. During the 1980s and 1990s, I used much time and energy writing and talking about what philosophy was, how philosophy could contribute to nursing and nursing research and what could not be expected from philosophy. The reason for this was that philosophy of nursing was a new area and many researchers, teachers and students using philosophical concepts were not trained in philosophy. A lot of misconceptions were around, and I saw it as my mission to spread some light on the connection between nursing and philosophy. Now this sub-discipline has developed nicely, much thanks to the nursing philosophy conferences in Canada and the UK and the international journal, *Nursing Philosophy*. Teaching ethics and research methods as well as supervising students writing PhD and Masters theses have also given opportunities to spread philosophical awareness.

The second line of activity has involved actually conducting philosophical studies related to nursing. This philosophical work has, to a large extent, been focusing on epistemology, i. e., forms of knowledge implicit in or in other ways related to nursing and nursing science. The starting point was my PhD thesis (Sarvimäki 1988a), where I outlined a view of the different forms of knowledge in interactive practice disciplines, such as health care, based on a synthesis of existentialism, pragmatism and constructionism – an eclectic endeavour I am not sure I would board today. Off-springs of this work can be seen in later articles where I offered a view of the discipline of nursing in terms of knowledge inherent in science and tradition (Sarvimäki 1994), and applied the forms of knowledge I had outlined previously in relation to moral or ethical knowledge (Sarvimäki 1995). The article on intuition mentioned previously (Sarvimäki & Stenbock-Hult 1996) was also an epistemological study.

In addition to the epistemological studies, I have dealt with questions related to health, well-being, quality of life, ethics and vulnerability. These topics have been more popular from the audiences' and readers' point of view, resulting in a large number of lectures, teaching sessions, articles in professional journals and text-books for students. Although these speeches and texts have mainly been presented to professional forums and the so-called general public, they have been clearly founded in philosophical ideas. An example of this is Stenbock-Hult's and my book on nursing ethics, published in Swedish in 2008 (Sarvimäki & Stenbock-Hult 2008), and the next year in Finnish, where we presented a comprehensive view on nursing ethics based on Heidegger's philosophy and an existentialist view of the vulnerability of human beings. A scholarly paper in this area is an article on well-being based on Heidegger's philosophy (Sarvimäki 2006).

My third line of activity involves the use of philosophical ideas and themes to influence empirical research. This research mainly includes studies of quality of life and the meaning of vulnerability that I have conducted in collaboration with Stenbock-Hult (Sarvimäki & Stenbock-Hult 2000; Stenbock-Hult & Sarvimäki 2011). In this research we have taken our philosophical points of departure in literature on the good life, meaning in life, being-in-the-world, the existential attitude – that is, themes in the existentialist tradition. They have formed a frame of reference that has inspired us to ask research questions and discuss results.

5. Where do you see the field of philosophy of nursing to be headed, including the prospects for progress regarding the issues you take to be most important?

Where is philosophy of nursing headed? Is it going anywhere? Where is philosophy headed? Where can it be headed? Can we see genuine progress in philosophy? How is progress in philosophy to be understood? What is the meaning of life? Why are we here? Questions, questions but no definite answers.

The only answer to the question "where is philosophy of nursing headed" is probably: forward. Having followed the field of philosophy of nursing during the past decades I can see that philosophizing has become deeper, more varied and richer in nuances. More philosophical. What surprises me sometimes, though, is that I come upon texts that deal with the same problems as philosophers of nursing did 20 – 25 years ago and much in the same way. Progress in the field does not exclude dealing with the same problems – in a way much of philosophy still occupies itself with problems formulated by Plato, Aristotle and other philosophers of antiquity. Progress in philosophy requires, however, that we get a deeper understanding of these problems, see them in a new light, re-formulate them on the basis of new scientific knowledge, technological development and social challenges.

In the same way as philosophy in general, philosophy of nursing will probably develop into separate traditions, some being analytically and conceptually oriented, some being close to the so-called continental philosophy and so on. In order for this to happen, there has to be enough scholars doing philosophy of nursing, forums for philosophical discussions and – above all – time. Philosophizing takes time and developing traditions takes time. Students, teachers and scholars need time to ask questions, reflect and engage in analytical discussions. And time is something that has become scarce in the production-marketing logic that has invaded nursing education and research as well as nursing practice.

In my answer to question 2 I called for more problematizing, questioning, digging, revealing and arguing. I cannot see why this could not happen in philosophy of nursing, if there are dedicated people who are given the opportunity and resources to engage in this sort of activity.

Philosophy of nursing can be characterized as an ongoing discourse between different views leading to greater clarity, deeper understanding – and new questions.

Selected Works

Kalkas, Hertta and Anneli Sarvimäki. 1985. *Hoitotyön eettiset perusteet*. [The Ethical Foundations of Nursing]. Helsinki: Sairaanhoitajien koulutussäätiö.

Sarvimäki, Anneli.

1988a."Knowledge in Interactive Practice Disciplines. An Analysis of Knowledge in Education and Health Care." *Research Bulletin 68*. Helsinki: University of Helsinki, Department of Education.

1988b."Nursing Care as a Moral, Practical, Communicative and Creative Activity." *Journal of Advanced Nursing* 13: 462 – 67.

1994."Science and Tradition in the Nursing Discipline. A Theoretical Analysis." *Scandinavian Journal of Caring Sciences* 8: 137-42.

1995."Aspects of Moral Knowledge in Nursing." *Scholarly Inquiry for Nursing Practice. An International Journal* 9: 343-53.

Sarvimäki, Anneli and Bettina Stenbock-Hult. 1996. "Intuition – A Problematic Concept in Nursing." *Scandinavian Journal of Caring Sciences* 10: 234-241.

Sarvimäki, Anneli. 1999. "Answering Philosophical Questions Facing Contemporary Nursing Practice." *Western Journal of Nursing Research* 21: 9-15.

Sarvimäki, Anneli and Bettina Stenbock-Hult. 2000. "Quality of Life in Old Age Described as a Sense of Well-being, Meaning, and Value." *Journal of Advanced Nursing* 32: 1025-33.

Sarvimäki, Anneli. 2006. "Well-being as Being Well – A Heideggerian Look at Well-being." *International Journal of Qualitative Research on Health and Well-Being* 1: 4 – 10.

Sarvimäki, Anneli and Bettina Stenbock-Hult. 2008. *Omvårdnadens etik. Sjuksköterskan och det moraliska rummet*. [Nursing Ethics. The Nurse and the Moral Space. In Swedish.] Stockholm: Liber.

Stenbock-Hult, Bettina and Anneli Sarvimäki. 2011. "The Meaning of Vulnerability to Nurses Caring for Older People." *Nursing Ethics* 18: 31 – 41.

21

P. Anne Scott

Professor of Nursing

Dublin City University, Republic of Ireland

---

1. How were you initially drawn to philosophical issues regarding nursing?

I went on duty one morning to the female surgical ward to which I was allocated. The tension in the nursing office was immediately clear. The well-respected night nurse began to give the "handover" report and moved rapidly to recount what for her was the most significant event of the night. Mrs X, an eighty-six year old woman, who had been admitted the previous afternoon suffering from an obstruction in her bowel, had "arrested" at 3 a. m. The medical staff had resuscitated her on three separate occasions over the following hour and she was now unconscious, but stable. The night nurse was outraged at the medical intervention, felt the elderly woman should not have been resuscitated even immediately following her initial arrest and should, given her age, have been left to "die in peace." In her view the second and third resuscitation was an affront to human dignity and she had strenuously opposed the medical staff decision both at the time of the intervention and now, some hours following the events recounted. This senior staff nurse had refused to look after Mrs X since the recent successful resuscitation attempt. She had told the medical staff that given their resuscitation interventions it was now their job to look after this patient – as what they had done went against this nurse's conscience and she would not be party to any further care of this patient. So influential was this staff nurse that the nursing staff who had come onto the "day shift," following the handover report, also refused to look after this patient. They told me, as the senior student, and the first year student who was also on placement on this ward, that we students could do as we liked.

At the time this incident rapidly set off a series of complex questions for me. From my perspective, even then, it was clear that it was problematic to "abandon" this patient just because one might object to a particular medical decision – in this case a series of resuscitation attempts, the last of which appeared to be successful. It was common among the nurses I had worked with to object to at least some medical decisions and or interactions – whether it was provision of insufficient information to patients or relatives as they saw it, inappropriate drug treatment or lack of or inappropriate referral – however nurses had never before, in my experience, refused to care for a patient. Making such a decision simply on the basis of an objection to a medical intervention was exceptional, peculiar and for me at the time deeply anxiety provoking. What was so different in this particular case?

At that time I was familiar with some of the issues that had arisen in the United Kingdom regarding abortion (still not, at the time of this incident or since, legally sanctioned in the Republic of Ireland). I was also aware or believed that although conscientious objection to participating in care was focused on relief from caring for a woman who had sought and/or had an abortion carried out, there had been unsuccessful attempts to extend the right to conscientious objection to caring for certain patients or groups of patients.

From my perspective the case we were now faced with was particularly distressing as the woman was not party to, and could not have been party to, the decision-making with regards to the medical intervention objected to by the staff nurse. The patient had suffered a cardiac arrest and the medical team intervened on three separate occasions to resuscitate her. The patient had remained unconscious throughout. Therefore why were we potentially "punishing" this defenseless, unconscious woman by not providing her with necessary nursing care, simply because we might object to the decision of the medical team?

This question led to a further set of questions: When should a patient be resuscitated or indeed not resuscitated? Who had the right to decide this? On what grounds might such a decision be made? Does the patient or her daughter, as next of kin, have a say in this?

The particular case also raised questions regarding what my responsibility was as a second year, thus senior, nursing student to a) the patient, b) my fellow junior student, and c) to the staff nurses with whom we both would have to continue to work for the remaining four weeks of placement?

I phoned the Principal Tutor, in the School of Nursing, for advice. He was very sympathetic but ultimately, if memory serves me correctly, the decision regarding what to do was left ambiguous. I do have a distinct recollection of feeling abandoned both by the clinical staff who normally provided input, guidance and feedback – sometimes unwelcomed – to my ward –based nursing practice. I felt similar isolation and lack of support from the person in the School of Nursing who had been, and for many years continued to be, a significant influence on my developing career.

Working through the particular situation on that particular day certainly posed challenges. However the case and the many questions it generated also remained with me for the remainder of my time as a student nurse and indeed continues to influence my thinking and teaching and practice both as a nurse and as a nursing academic.

In 1979, also during my second year of nurse training, there was significant unrest among the nursing profession in Ireland; largely around terms and conditions of nursing employment. Finally, in my training hospital as a result of very significant overcrowding in the hospital, there was a nurses' strike. Emergency cover was maintained but all other nursing input and activity was severely restricted. There was much discussion at the time regarding the morality of nurses who strike – who cares for vulnerable patients, do nurses actually have a right to strike? How and who should decide on this issue? What would be the implications of deciding that nurses did not have the right to strike? The experience of that period in the hospital, and the distressing conditions experienced by both patients, who were nursed in beds placed up the centre of Nightingale wards and around hospital corridors, and the nurses who cared for them crystallised the need to be able to draw limits and boundaries to acceptable behaviours and acceptable care. For many years, while struggling to answer such questions, I refused to become a member of the Royal College of Nursing (RCN), on the basis of the RCN anti-strike clause.

A further and pervasive issue for me as a student nurse, and as a young staff nurse and ward sister, was something along the following lines: "What is the core of nursing?" What is the correct balance between any required professional distance and humane, sympathetic, engaged, compassionate care for a particular patient in a particular set of circumstance? Is it possible to be too engaged, too involved? What is the balance between the supportive, coaching, prompting role of the nurse and an authoritarian power trip where a vulnerable, compromised patient, who has suffered a cardio-vascular accident for example, is "forced" to try again and again to relearn to feed, dress and care for him/herself?

How does the nurse avoid institutionalisation and the destruction of normal social and professional standards particularly when "caring" for very cognitively compromised or disempowered clients? A particularly mind-focusing example of this arose on my 6 week "psychiatric nursing" placement. I was rostered for weekend duty on a long stay male unit. On the Saturday morning, following the "report" and a cup of tea with the day staff, I was told that this was "bath day". This meant that a number of the patients would need to have their weekly bath prior to lunch. I was given the names of the particular patients and told that nurse X would assist me. I went off to let the relevant patients know they were to have their baths before lunch and then went in search of towels, shampoo and so forth. I was completely astounded to arrive back at the bathroom some ten or so minutes later to see five naked men lined up outside the bathroom door. Each was then steered in turn to wash themselves in the same bath water that their fellow patients had just used. What enabled amiable, approachable, apparently competent staff to think that this was an acceptable way to treat their patients?

There were and are of course many other complex and interesting issues with potentially significant implications for both patients and staff. Admissions and work on surgical wards, in particular, frequently seemed to raise resource allocation questions – most often in terms of dividing one's time among what at least appeared to be equally needy and vulnerable patients. The issues of scarcity, rationing and resource allocation took on a much heightened focus when I moved to work as a volunteer nurse in a bush hospital in North West Kenya in the mid and late 1980s.

Practice as a student of nursing and as a qualified practitioner is, in my view, a potential hot bed of philosophical and ethics questions. I struggled with many of these questions as a student nurse and only later, as an undergraduate in Philosophy, was I properly equipped to categorise and fully articulate some of the questions; questions which for me remain core to nursing and the nature of nursing practice.

2. What, in your view, are the most interesting, important, or pressing problems in contemporary philosophy of nursing?

From my perspective some of the key issues in contemporary philosophy of nursing are the following:

1) The articulation of the nature of nursing as a practice in health services of the 21st century– what is the core of nursing – its core values, characteristics and is this as it should be? What are the criteria to determine appropriate answers to these questions? What is the moral imperative of nursing as a practice?

2) Issues of rationing and allocation of the limited nursing resource. The current climate of "austerity" in many western health systems brings these issues into sharp relief.

3) What should the balance be between supporting and caring for this particular patient/group of patients who currently present as being in need, and the need to draw limits in terms of what can reasonably be expected from the professional nurse? Should strike

action be available to nurses as an ultimate sanction – in my view the answer is yes, but within clearly delineated boundaries. If one cannot ultimately draw one's own boundaries, one ultimately cannot call a halt when specific elements become particularly problematic or apparently intractable – thus one is like "a candle blowing in the wind." The provision of poor, potentially unsafe care, for whatever reason – severely reduced resources included – may be worse than providing no professional care at all. This idea needs to be aired, discussed and recognised. In a "no care" situation, for example, patients and relatives/carer where they exist, are aware that there is no professional care being provided and thus do their best under the particular circumstances. In the poor care situation patients and relatives/carers may be under the illusion that care (i. e. appropriate care) is being provided. Thus patient/carer credulity may be present/heightened and vigilance, attention and advocacy may be lacking – thus reducing basic survival mechanisms.

3. What, if any, practical and / or socio-political obligations follow from studying nursing from a philosophical perspective?

Consideration of any practice, nursing included, from a philosophical perspective gives one a greater understanding of the conceptual base upon which the practice rests. Philosophical study helps identify and distinguish empirical and philosophical/analytical questions and, on occasion, shows where answers to each may compliment the other. It also helps elucidate the core values of the discipline. Questions such as "Is nursing an art, a science or a practice – or has nursing elements of all three?" and "What does caring within the context of nursing practice mean?" are the kinds of questions that are interesting from both a conceptual and a practical perspective.

The practical implications of deliberating on such questions is that in the first instance one becomes clearer regarding, for example, the core concepts and values underpinning and/or permeating the discipline and practice of nursing. Secondly such clarity should inform one's practice; how one interacts with patients, engages with colleagues, educates and mentors students, and how one presents one's case for a share of the health service resource, in order to provide appropriate nursing care for the patients in one's charge.

This latter issue, working to obtain an adequate share of the health service resource in order to provide appropriate care for one's patients, is, in my opinion, a key issue facing nurse leaders and managers in health services of the 21st century. However, in my view, it is also a task that, in general, nurses – including nurse leaders and managers – are not well equipped to tackle and one which the traditional roles, expectations, gender profile of nursing in society, in health care delivery and in health service hierarchy makes particularly difficult to tackle. Therefore from my perspective this task is one which nursing academics, particularly those who have a grasp of both the conceptual/philosophical issues and the empirical evidence available, have an obligation to work on with nurse leaders both locally and nationally. We should work collectively to develop coherent, evidenced, well argued cases to be used with health policy developers and decision makers, in order to seek adequate resources to facilitate nurses in providing the type of patient care that our societies claim to desire for the ill and vulnerable.

In my experience it is frequently suggested that one of the difficulties for nursing is that we are unclear on the nursing contribution to patient care and patient outcomes, and this makes it exceptionally difficult to argue nursing's case effectively with health ministries and health budget holders at community, hospital, regional and national levels. However I would argue that the claim that the nursing contribution to patient care and patient outcomes is unclear is at best flawed and at worst simply inaccurate. Clarity on the meaning of 'patient care' and 'patient outcomes' of course needs to be explicit in any such discussion. However taking this as read (and momentarily ignoring the work of founders of modern nursing in Europe and the USA such as Florence Nightingale, Dorothea Dix, Mary Seacole, Edith Cavel, Mary Todd Lincon, Mother Mary Francis Bridgeman, Catherine McAuley, Mary Aikenhead and Irena Sendler), there is now at least 20 years of international literature evidencing the nursing contribution to both patient care and patient outcomes (see for example Aiken et al 2002, 2003, 2008, 2011, 2012; Benner and Wrubel 1989; Buller and Butterworth 2001; Care Quality Commission 2011; Francis 2010; Institute of Medicine 2004, 2011; Jinks and Hope 2000; Kane 2012; Moran 2008; Papastravrou et al 2011; Rafferty et al 2007; Scott et al 2006). Combine the above with the significant literature in nursing philosophy (including nursing ethics) and I suggest that the contribution of nursing to patient care and patient outcomes can be distilled down to two key contributions: nursing is elemental for safe, quality care provision and nursing is important for the humane, compassionate treatment of patients. Among the questions that now require answering are the following. What type of nurse education best ensures such patient outcomes (Scott 2007)? How many nurses are required to provide such nursing care – i. e. what density of nurses in a population, what nurse-patient ratios in the variety of in-patient and community settings in which nurses of the 21st century work and in which patients are cared for? Do nurse education levels matter in this scenario? Is it possible to get more effective patient outcomes by attending to work environments and organisational culture rather than simply focusing on nursing numbers alone? How do and how should societies and governments balance the needs of health care, social care, housing, education and so forth. Clearly there is a mix of empirical and analytical questions to be addressed.

However the practical and socio-political obligations emanating from the knowledge we already have at our fingertips should be reasonably clear to any group, organisation or society that takes or claims to take caring for the vulnerable and the ill seriously. Articulating and accepting these conclusions would go some way to achieving a shared understanding of the role and function and place of nursing in a 21st century health system. Such shared understanding seems important in order to provide the foundations for the necessary discussions regarding resource allocation, including staff patient ratios, and so forth.

As those of us who study nursing from a philosophical perspective will appreciate becoming clear and coming to know what is required or what should be done is important. However ensuring that the appropriate action is taken is another matter. From my perspective there is an obligation on those of us "in the know" to work with those leaders who can use such knowledge, to do all within our collective power to both explain nursing and explicate its importance to good quality, safe and humane patient care. As Mukens argued many years ago nurses are not passive pawns in the health system or in the provision of inadequate care and service to patients, they are often colluders in such service and such care – some might suggest that recent cases such as Mid Staffordshire (Francis 2010) and the Care Quality Commission report into facilities where we care for elderly people (CQC 2011) are cases in point. However without the human resources necessary to provide safe, humane care nurses cannot reasonably be held accountable for the provision of inadequate care. This is a conclusion that is controversial. It is dangerous for patients and it poses many challenges for the nursing profession.

Accountability implies autonomy and choice. If there is no alternative to inadequate care, due to severe understaffing or lack of other required resources/infrastructure, then choice does not exist. Thus in order for patients, employers, nurse leaders and policy makers to reasonably demand that nurses "step up to the plate" they must ensure the resources, culture and climate that facilitates such accountability.

Nurses and nursing leadership must be held accountable but only for actions they can choose, influence and control. As academics interested in nursing philosophy, as tax payers, as voting members of the public and as potential future patients and recipients of health care we must also be held accountable and must try to hold government, health policy makers and health service leaders accountable for placing patients and nurses in untenable care delivery positions, when such is an accurate description of the actual situation patients and staff find themselves in. This is a more honest appraisal and likely to ultimately lead to a better outcome for patients and nurses than the familiar attempts to discipline the "bad nurse" (see for example Kellett 1996a, b) or identify the "rotten apple".

4. In what ways does your work seek to contribute to the philosophy of nursing?

My own work seeks to identify, articulate, extend and clarify the conceptual basis of the discipline of nursing, largely as a necessary precursor to developing both a theoretical and a practical understanding of nursing as discipline, profession and practice with a significant moral component. For me questions in health care ethics are intimately connected with conceptions of personhood, autonomy, nursing (including that character of the nurse), medicine (and the character of the doctor) and patient (including questions of rights and responsibilities). A theoretical understanding of, for example, nursing practice and the character of the nurse is at least a co-requisite to informed discussion on models of nurse-patient interaction, the caring role of the nurse, conceptions of patient dependence/autonomy, notions of protective responsibility (Holm 1997) and so forth. Articulating, discussing and challenging certain conceptions of autonomy is relevant in both considering how one balances patient autonomy and desires with professional responsibility; and with considering the moral agency of the nurse within the context of a variety of practice contexts. Understanding the components of the conceptual, theoretical and social frameworks within which one is operating also is required to inform discussions on the limits of what is reasonably demanded of the individual nurse within the context of a particular practice environment/patient care scenario.

5. Where do you see the field of philosophy of nursing to be headed, including the prospects for progress regarding the issues you take to be most important?

Significant foundation work has been done in the initial mapping of the terrain. However philosophy of nursing is a neophyte in the history of ideas. It is largely a product of last 40 years or so, largely initiated in USA with work on conceptual nursing models. Interest in ethical issues in medicine, health care and nursing is largely a development of the 20th century, with seminal contributions again initially from the USA, then UK and Scandinavia.

Gradually there are increasing numbers of philosophers who are interested in health care, including nursing, from a philosophical and ethical perspective. There are also gradually increasing numbers of nurses who have studied philosophy/ethics. This strengthens the reach and depth of the philosophy of nursing discourse and can lead to rich interdisciplinary discourse that ultimately may assist disciplinary development and enhance practice.

As one who, over the past 12 years, has contributed to the marginal rather than core discourse I think there is a need to intensify the theoretical discussion while mapping, clearly, the practical implications. Philosophers of nursing must also, in my view, be prepared to grasp and use empirical work to support elements of their philosophical work where it is relevant to do so. There is a growing empirical literature that is relevant to conceptual discussions regarding patient need, desire and comfort with nursing; and to conceptualising the focus and limits of nursing practice in the 21st century. For me this has, and continues to be, a hugely challenging, interesting and intellectually stimulating field of activity.

References

Aiken LH, SP Clarke, DM Sloane, J Sochalski and JH Silber. 2002. "Hospital Nurse Staffing and Patient Mortality, Nurse Burnout and Job Dissatisfaction." *Journal of the American Medical Association* 288: 1987-1993.

Aiken LH, SP Clarke, RB Chen, DM Sloane and JH Silber. 2003. "Education Levels of Hospital Nurses and Surgical Patients' Mortality." *Journal of the American Medical Association* 290: 1617-1623.

Aiken LH, J Buchan, J Ball and AM Rafferty. 2008. "Transformative Impact of Magnet Designation: England Case Study." *Journal of Clinical Nursing* 17: 3330-3337.

Aiken LH, J Cimiotti, DM Sloane, HL Smith, L Flynn and D Neff. 2011. "The Effects of Nurse Staffing and Nurse Education on Patient Deaths in Hospitals with Different Nurse Work Environments." *Medical Care* 49: 1047-53.

Aiken LH, W Sermeus, K Van den Heede, DM Sloane, R Busse, M McKee, L Bruyneel, HL Smith, A Kutney-Lee, AM Rafferty, P Griffiths, MT Moreno-Casbas, C Tishelman, PA Scott, T Brzostek, J Kinnunen, R Schwendimann, M Heinen, D Zikos and IS Sjetne. 2012. "Impact of Nursing on Patient Safety, Satisfaction and Quality of Hospital Care in 12 countries in Europe and the United States." *British Medical Journal* 344: e1717 *(http://www. bmj. com/content/344/bmj. e1717)*.

Benner P and J Wrubel. 1989. *The Primacy of Caring: Stress and Coping in Health and Illness*. Menlo Park, CA: Addison-Wesley Publishing Co.

Buller S and T Butterworth. 2001. "Skilled Nursing Practice – A Qualitative Study of the Elements of Nursing." *International Journal of Nursing Studies*, 38(4): 405-417.

Care Quality Commission (2011) *Dignity and Nutrition Inspection Programme: National Overview.*
(http://www. cqc. org. uk/sites/default/files/media/documents/
20111007_dignity_and_nutrition_inspection_report_final_update. pdf. accessed 2nd August 2012).

Francis R. 2010. *Independent Inquiry into Care Provided by Mid Staffordshire NHS Foundation Trust January 2005 – March 2009, (Vol 1)*. Chaired by Robert Francis QC. Stationary Office, London.

Holm S. 1997. *Ethical Problems in Clinical Practice*. Manchester, England: Manchester University Press.

Institute of Medicine (IoM). 2004. *Keeping Patients Safe: Transforming the Work Environment of Nurses*. Washington, D. C.: The National Academies Press.

Institute of Medicine. 2011. *The Future of Nursing: Leading Change, Advancing Health*. Committee on the Robert Wood Johnson Foundation Initiative on the future of Nursing at the Institute of Medicine. The National Academies Press, Washington, D. C. ISBN: 0-309-15824-9

Jinks AM and P Hope. 2000. "What Do Nurses Do? An Observational Survey of the Activities of Nurses on Acute Surgical and Rehabilitation Wards." *Journal of Nursing Management* 8: 273-279.

Kane R. 2012. *A Case Study of Emergency Department Practice: Re-framing the Care of People Who Self-harm*. Unpublished PhD thesis, Dublin City University, Dublin, Ireland.

Kellett J. 1996a. "Taking the Blame" [with a commentary by Moore S]. *Nursing Standard,* 11(12): 21–23.

Kellett J. 1996b. "An Ethical Dilemma: A Nurse Suspended." *British Medical Journal* 313: 1249–51.

Moran A. 2008. *The Lived Experience of Haemodialysis Patients: Treatment and Quality of Life*. Unpublished PhD thesis, Dublin City University. Dublin.

Papastavrou E, G Efstathiou and A Charalambous. 2011. "Nurses and Patients Perceptions of Caring Behaviours: Quantitative Systematic Review of Comparative Studies. *Journal of Advanced Nursing* doi: 10. 1111/j. 1365-2648. 2010. 05580. x

Rafferty AM, SP Clarke, J Coles, J Ball, P James, M McKee and LH Aiken. 2007. "Outcomes of Variation in Hospital Nurse Staffing in English Hospitals: Cross-sectional Analysis of Survey Data and Discharge Records." *International Journal of Nursing Studies* 44: 175-182.

Scott PA, MP Treacy, P MacNeela, A Hyde, R Morris, A Byrne, M Butler, J Drennan, P Henry, M Corbally, K Irving and G Clinton. 2006. *Report of a Delphi Study of Irish Nurses to Articulate the Core Elements of Nursing Care in Ireland*. Dublin City University. Dublin. ISBN 1872327605

Scott PA. 2007. "Nursing and the Notion of Virtue as a 'Regulative Ideal'." In *The Philosophy of Nurse Education* edited by JS Drummond and P Standish, 33-45. Houndmills, Basingstoke: Palgrave Macmillan.

22

Derek Sellman

Associate Professor and Director of the unit for Philosophical Nursing Research

University of Alberta, Edmonton, Canada

---

1. How were you initially drawn to philosophical issues regarding nursing?

I don't think mine a particularly unusual story insofar as I came to nursing after an inauspicious start to a working life that encompassed a variety of jobs well suited to my distinct lack of qualifications on leaving school. After some 5 years of insignificant if varied employment it was no more than serendipity that led me to a 3-year programme of psychiatric nurse training rather than the 2-year version with all its associated career limitations (as was the fashion in those times, I was later to complete a shortened general nurse training). And serendipity seems to loom large in the story of how I became associated with the philosophy of nursing. I suppose it is fair to say that as far as I can recall the tendency to what Emmet (1991) would understand as philosophizing was one of my enduring, but not always endearing, traits. Of course, it was only much later in life that I came to understand philosophizing as my default position although it might be that I have overplayed this tendency as a way of explaining my failure to thrive in the British school system of the 1960s. I was always curious but not, it would appear, about the things on the curriculum. Further, the didactic modus operandi of Chase Cross Secondary Modern School for Boys did little to satisfy my curiosity and it is, I think, this restless quest to understand that draws me towards philosophical thinking; and it is because I am a nurse that I am drawn to philosophy of nursing.

Nursing was my epiphany insofar as this was where, over time, I discovered a love of learning that lead me to a life of formal part-time study that progressed from a single 'O' level in 1975 to PhD in 2005. Along the way I have to thank a number of individuals who recognized in me things I did not recognize in myself. In terms of my arrival at all things nursing philosophy there are five or six prominent individuals implicated. The first was the head of a school of nursing who informed me on my return from vacation one year that I was to lead the *Philosophy and Ethics* theme in the new Project 2000 curriculum.[1] It was Hobson's choice, of course, although the answer to the 'why me?' question was a flattering 'in this school you are probably the person best suited to the theme'. I took it as a compliment. And so it was that after about six weeks of preparation I began teaching nursing ethics to a first cohort of Project 2000 students. To them, and to several subsequent cohorts, I should probably now take this opportunity to apologize for my lamentable lack of formal philosophical training and for my minimal competence in teaching a subject about which I could not at that time lay claim to any particular expertise. As the saying goes: 'you never really know what it is you know until you try to teach it' – and, boy, did this hit home in those first few classes. In the attempt at a remedy, there quickly followed a Master's in Teaching Heath Care Ethics at the Institute of Education, University of London led by Graham Hayden. This Master's programme certainly helped direct the teaching required of me but perhaps more importantly, gave me access to a language and a vocabulary with which to begin to give expression to my thoughts – a language with which under the sympathetic supervision of Patricia White gave me the confidence to pursue a PhD in Philosophy of Education. Philosophy of Education shares several significant features with Philosophy of Nursing, not least of which is the practical nature of their respective core activities. With yet another serendipitous flourish it was a chance encounter with Keith Cash who, as he moderated a conference session at which I gave a concurrent paper, encouraged me to head west for the first Swansea International Philosophy of Nursing conference organized by Steven Edwards and Paul Wainwright. The rest, as they say, is history. I still sometimes feel that my lack of formal training in philosophy hinders my contribution but more on this later.

2. What, in your view, are the most interesting, important, or pressing problems in contemporary philosophy of nursing?

One thing that my lack of a standard academic background has given me is a somewhat eclectic interest in more things than I can possibly attend to in one lifetime. It has also left me effectively outside of any

[1] In the UK it was the Project 2000 reformation of nurse education that, amongst other things, changed the status of those learning to become nurses from apprentices to bona fide students and was a precursor to the wholesale move of nurse education into universities and other institutions of higher education. It also saw the curriculum developed on the lines of subject themes which in turn lead to the requirement for nurse teachers to move from generalists (teaching whatever topic was on the timetable) to specialists (teaching to, and becoming 'expert' in, a particular theme and its application to nursing).

strict adherence or loyalty to any one discipline or any one tradition within a discipline – for I have never suffered the fate of being trained in a particular school of thought or methodology, rather during most of my formal educational programmes of study I have been encouraged to think for myself outside of any the formal constraints to which others I have known have been required to conform. I do not know if this eclecticism is a good thing but it does serve to make me highly suspicious of strong (normative or imperative) claims about nursing, particularly where these claims rest on assertion rather than argument. It also makes me highly suspicious of what might be termed those 'fashionable movements' generated by the seemingly whimsical adoption of sometimes politicized ideas, or of particular theorists or philosophers that tumble out of, amongst other places, the pepper pot of nursing's movers and shakers. Foucault, no doubt, would have something to say about the reasons why some rather than other of these movements capture the imagination of either nurses in general or of those with influence in what is to become for nursing 'this year's black'. For me the pressing philosophical problem lies not with the adoption of *this* rather than *that* idea, ideology, or idol, but the way in which nurses allow these movements to become *de rigueur* in the absence (largely) of the kind of critical debate that might curtail their sometimes evangelical (by some) and passive (by others) acceptance. I shall refer to this as the 'orthodoxy obsession'.[2]

This orthodoxy obsession, which I take to be neither solely contemporary nor restricted to nursing, presents us with a set of important and pressing problems insofar as it fragments the narratives and traditions that MacIntyre (2009) indicates as necessary for practices to flourish in the world. It is not merely a case of the triumph of austerity-driven rampant managerialism (although that this has corrosive effects on the work that nurses do is not denied) rather it is that nursing's response to the conditions within which nursing practice occurs reflects the same internal and external pressures that maintain the impetus for manage-

---

[2] By using the term 'orthodoxy obsession' I mean to capture the seemingly constant tendency of nursing to seek to reinvent itself by the hasty and usually uncritical adoption of what often turns out to be a 'fad' of one sort or another dressed up as the answer to all nursing's problems only to be replaced subsequently by the next panacea. Practitioners will understand this in terms of the never-ending stream of yet more demands for changing the way things are done. Indeed, practitioners might feel that they cannot turn around for fear of bumping into an edict that tells them that what they are in the middle of doing is now no longer permitted – be it a practical or paperwork task. I'm not sure how much of this actually effects the everyday work of most nurses as one danger of this ever-changing environment is that nurses become immune to the constant directive declarations and, wherever possible, merely carry on as before.

rialism to dominate. The continuing attempts at the reinvention of nursing witnessed by both the frequent replacement of one orthodoxy with another and by the existence of competing and sometimes incommensurable orthodoxies is something with which nursing is complicit in generating as well as imposing orthodoxies of uncertain and perhaps unexamined provenance. I suspect the critical theorists are correct in noting that we are the victims as well as the authors of the doxy that lies behind our understandings of the world around us but would that they apply the same critique to their own traditions and assumptions.

3. What, if any, practical and/or socio-political obligations follow from studying nursing from a philosophical perspective?

I'm not sure how to go about addressing this question. On the one hand I recognize the strengths of the argument that tells us cognition without action plays into the hands of those who wish to retain the status quo – to the point that at times I am almost convinced of the correctness of this claim. But then, on the other hand, my default skepticism kicks in and I become suspicious of the grand claim that lies within, particularly when that claim takes a limited binary form. I think, in part, this has something to do with the tendency of the argument for action to fall back on the assumption that nursing cannot be divorced from the political which, while undoubtedly true, is contingent not only on how the political is understood but also on how nursing is conceptualized – this latter, of course, is one of the those unresolved questions that get aired from time to time particularly when some event or scandal occasions questions about the purpose of nursing. Moreover, as a statement of intent, merely noting that nursing is political does little, as Lipscomb (2011) notes, to address the practical implications of how this is to translate into action. Thus the oft assumed obligation of activism that some think studying nursing from a philosophical perspective requires is itself very likely to be a function of a particular ontological perspective; and without denying that there are things in the world of nursing that ought to be other than they are currently are, it is not clear that nurses (either collectively or individually) should engage in socio-political action of the type that some exhort. This is not to say that nurses (again, either individually or collectively) should not be active socio-politically but it is to suggest that there would seem no *de facto* requirement for nurses *qua* nurses to act on such obligations as some would claim to arise from studying nursing from a philosophical perspective.

As an alternative, and from a particular standpoint, such obligations as may arise might not be so much those that follow the study of nursing from a philosophical perspective but instead, those that necessitate nurses philosophize (in the sense that Emmet uses that term) about nursing. This is perhaps a demand that nurses do no more than think about what it is they do, what they do not do, and what they might otherwise do; to think about what nursing is, what it is not, and what it might otherwise be; and so on. These things seem to be implied in any conception of thinking philosophically about nursing and would seem to be within the grasp of any nurse. As Emmet writes:

> Too often philosophy tends to be regarded as a remote and abstruse subject which can only be profitably studied by the brilliant few. It seems to me that this is unfortunate and that philosophical matters are often less difficult and more important than is generally supposed. We all philosophize whenever we attempt to handle abstract ideas and it may matter very much whether we do it well or badly. (1991, 9)

And one might easily add 'nursing' in appropriate places within this statement of Emmet's and so give expression to the idea of the importance of thinking about nursing in the abstract as well as in the particular. The obligation here then would be to philosophize in ways that offers insights into the phenomenon that is nursing. Such philosophizing seems to me to require a critical openness to ideas, lines of thought, and claims that may be challenging, contradictory, or controversial – and it would be of benefit to our understanding of nursing that such pursuit be conducted not from some perverse desire to be deliberately obtuse or contentious, and not merely in the attempt to persuade others of the rightness of our deliberations, but for the genuine purpose of discovery. I suspect that any obligations that surface during or after such philosophizing will very much depend upon the conclusions (tentative or otherwise) of this philosophizing. In other words, on this view any prediction about what obligations might arise from studying nursing from a philosophical perspective would seem to presuppose the outcome or result of such study. And this seems somehow inconsistent with the kind of open enquiry that I take to be constitutive of philosophizing well.

4. In what ways does your work seek to contribute to philosophy of nursing?

At the risk of overstating what should by now be fairly obvious to the reader, philosophizing is part of who I am. I can imagine no other way of being, so when asked how my work seeks to contribute to philosophy of nursing I struggle to come up with a convincing answer. I suppose on a generous account I might say that in my work I seek to inspire inquisitiveness in others about nursing's taken for granted. And this would be closely associated with seeking logic in nursing education taking the form of, for example, if *such and such* is true then whatever educational implications follow from this requires us to adopt appropriate changes in the curriculum. In a way this might have been something I could have suggested in response to the previous question as an obligation that follows from studying nursing from a philosophical perspective but given the tentative and contestable nature of philosophical pursuits there might be some merit in not acting too hastily particularly if next year *such and such* turns out to be last year's black. But in thinking philosophically about nursing it seems to me there is much that has relevance to learning to become and to remain a nurse beyond mastery of technical skill. In continuing to write from this perspective, in which task I am merely one of many voices, I suppose it would be fair to say that I seek, like many others, to influence the way nursing is perceived, particularly by nurses themselves. Taking as a starting point Schön's (1983) critique of the limits of technical rationality and its failure to explain adequately expertness in professional practice, I seek to remind colleagues that even when nursing is understood as just another job of work the simple 'off the shelf application of protocol' approach rarely satisfies either the complexity of much that encompasses nursing or the human desire for care. Of course, whether or not nurses or nursing *should* care is an ontological question that periodically exercises the minds of nursing scholars. But leaving aside the care question, the technical rationality of which Schön warned us is writ large in the form of the modern protocol that dictates much of the work that nurses do. This tendency to the protocolization of work generally and of nursing work in particular seems to have been given greater impetus by the retreating economic prosperity to which we had become accustomed during the second half of the 20th century. It is of concern, I think, that many of the protocols that bear upon the work of nurses emerge from a close alignment of institutional, organizational, political, ideological, and even professional concerns which converge to militate against the expression of professional judgment at the level of the individual nurse. The deleterious effect of this on practice can be seen where, for example, deviation from protocol is discouraged by financial or other sanctions and under these circumstances it is difficult to see how claims of professional autonomy and accountability can be sustained. If it hasn't become so already, protocol seems well on the way to becoming the type of ritual from which the evidence-based practice movement promised liberation.

On this view, protocol replaces ritual as task-based nursing work returns as the dominant form of practice despite the purported advances of the last half century or so (I suspect that in some places task-based nursing never really went away). Far from being autonomous, nurses become automatons adept at applying protocols to fit (sometimes nursing) diagnostic categories evaluated by industrialized models of organization with such things as throughput targets and league tables being evermore present as measures of success or failure. In this dystopian future characteristics that obstruct the efficient operation of health care services are systematically discouraged by nothing so much as the stress that arises from admitting to such characteristics and by the sheer volume of work required of any one individual during a working shift. Goodbye compassion, goodbye discretion, goodbye professional accountability as anything other than blind obedience to protocol.

Alternatively, and at the risk of simplistic binary thinking, protocol becomes the guide for practice and deviation is actively encouraged in the exercise of professional judgment where justificatory demands shift away from those who have professional reason to deviate from the protocol to those who have ideological capital invested in its generalized application. Nurses, and other healthcare professionals, are rewarded for standing against institutionalized protocols that fail to account for the particular circumstances of individual patients; and the maintenance and development of those characteristics that allow nurses to remain sensitive to the human cost of sickness are positively encouraged. These characteristics are consistent with on the one hand professional and lay conceptions of nursing recognized as something more than mere task oriented work and on the other, in the various if not always explicit claims found in provincial, regional, national, and international nursing codes. And it is in writing about these characteristics that I seek to contribute to an understanding of nursing as work of a particular kind. Work, as Sockett (1993) would say, that seeks human betterment as one of its guiding principles. That the characteristics of which I speak are compromised by the modern condition in which, at least for those of us in employment, there is always more to do than can possibly be done in the time allocated; always more directives seeking to control more of what we do; always another protocol to add to the growing folders of protocols on our desktops; and always another new (usually electronic and unforgiving) form for us to complete to account for how we spend the company's time. Yet, the human spirit seems not so malleable that it will put up with this indefinitely. Perhaps there is some room for optimism in the idea that there will come a point at which a popular reappraisal of the institutionalized influences that impose increasingly insidious ideologies to further objectify and commodify human experience will lead to an active (rather than rhetorical) reaffirmation of the human values important to the practice of nursing. In the meantime I will continue in my attempt to write in ways that encourage open dialogue in and about nursing.

5. Where do you see the field of philosophy of nursing to be headed, including the prospects for progress regarding the issues you take to be most important?

If Emmet is right in suggesting that "We all philosophize whenever we attempt to handle abstract ideas" (1991, 9) then it seems that the field of philosophy of nursing is alive and well for there is a strand of nursing scholarship that concerns itself very much with abstract ideas. I do not know how many authors of the scholarship of ideas in nursing would understand or call their work philosophical despite its obvious contribution both to the way others think about nursing and to the continuation of an enduring thread of philosophizing about nursing that can be traced back to at least Nightingale. The present and increasingly instrumental emphasis on research in nursing does pose a challenge for thinking about nursing in abstract ways and not only because it tends to squeeze the time and space that might otherwise be available for cogitation and consideration. As Emmet continues "...it may matter very much whether we do it [philosophizing] well or badly" (9) and while it may well be possible to philosophize well even under the pressure to publish, it might also be that this pressure curtails or constrains the activity of philosophizing. In the managerialist present those who pursue work in nursing of a philosophical persuasion might be perceived as engaged in trivial rather than fundamental pursuits. But despite these challenges I would want to remain optimistic although in terms of progress, I'm not really sure that I have a sense of how progress in philosophy of nursing might be measured. Looking back over my responses to these 5 questions, I suppose I would want to say that as long as there continues to be individuals willing to think and write about abstract ideas in nursing and as long as nursing itself can avoid falling prey to one or other fundamentalist orthodoxy that disallows critical engagement then the field of philosophy of nursing will be doing okay.

References

Emmet E. R. 1991. *Learning to Philosophize*. Harmondsworth: Penguin. (First published by Longmans in 1964; revised and published by Pelican 1968).

Lipscomb, M. 2011. "Challenging the Coherence of Social Justice as a Shared Nursing Value." *Nursing Philosophy* 12: 4-11.

MacIntyre, A. 2009. *After Virtue: A Study in Moral Theory* (3rd ed). Notre Dame, IN. University of Notre Dame Press.

Schön, D. A. 1983. *The Reflective Practitioner: How Professionals Think in Action*. Aldershot, England: Ashgate Publishing.

Sockett, H. 1993. *The Moral Base for Teacher Professionalism*. NewYork: Teachers College Press.

Selected Works

Sellman, D. 2007. "On Being of Good Character: Nurse Education and the Assessment of Good Character." *Nurse Education Today* 27: 762-767.

2009. "Practical Wisdom in Health and Social Care: Teaching for Professional Phronesis." *Learning in Health and Social Care* 8(2): 84-91.

2010. "Values and Ethics in Interprofessional Working." In *Understanding Interprofessional Working in Health and Social Care: Theory and Practice edited by* K. C. Pollard, J. Thomas and M. Miers. Basingstoke, UK: Palgrave Macmillan.

2011. "*What Makes a Good Nurse: Why the Virtues are Important for Nurses*". London, UK: Jessica Kingsley.

2012. "Reclaiming Competence for Professional Phronesis." In *Phronesis as Professional Knowledge edited by* E. A. Kinsella and A. Pitman, 115-130. Rotterdam, The Netherlands: Sense Publishers.

23

Sally Thorne

Professor, School of Nursing

University of British Columbia, Vancouver, Canada

---

1. How were you initially drawn to philosophical issues regarding nursing?

I suppose the initial draw toward philosophy for me was the intrigue of the search for meaning in illness, seeking to understand the existential components of the engagements we have with other human beings in the course of our work. Like most of us, I would not have articulated this as a philosophical kind of problem in the early stages of my career – a conceptual or ethical problem perhaps – until I began to engage with the qualitative health research movement in the early 1980s. The idea that a profoundly different paradigm of understanding truth might co-exist with science, and the possibility that these two sets of apparently mutually exclusive ideas might each play a role in our discipline, intrigued me. I saw a world of scholarly positionality around me in which both science and theorizing were necessarily firmly located. I wondered if I too ought to be taking a position or whether that might in fact be a limiting move. Although I was highly attracted to a wide range of "isms" at different points in my intellectual development, I found I could never devote myself entirely to any of them, as each seemed to have its strengths and limitations. And so I increasingly became intrigued by the idea of ideas.

Another draw for me was the opportunity to teach graduate courses in critical thinking and conceptualization in nursing. So often, we teach those two operations separately, and I think they create an interesting challenge when aligned into the same conversation. While the earlier courses I may have taught would have separated them somewhat, they tended to be courses I taught in sequence and with the same sets of students, and students most certainly fuelled a continually evolving set of inquiries about how ideas worked.

I believe that concepts and conceptualization serve as the foundation of our thinking, as they allow us to communicate, imagine, measure, and interpret. However, I do not see the essential challenge for nursing as operationalizing them for the purpose of measurement (Thorne 2005). In fact, what I learned from a generation of graduate students who would set out on the path of exploring the operational potential of core nursing concepts for the purposes of a graduate level assignment was that the better the critical thinking, the further that analysis took them from a comfort with measurement. Rather, many of the available measures started to seem quite hollow and vapid in comparison to the dazzling insights that were forthcoming from a broader, more abstract and multidimensional definition of important ideas. Spirituality, for example, is an idea that can become incredibly codified by definitional politics. And yet nurses generally recognize the more abstract idea that there is an element of human health experience that exists in the inarticulable and ineffable, often just beyond our grasp. Even our best attempts to capture this idea in words seem to miss the essence that is its nature. And so for me, the project of trying to "nail down" some of these gelatinous and amorphous ideas simply created a sterile and quite hollow ideational world in which they lost the meaning potential they might have held for us.

Philosophy, of course, came to the rescue in that it allowed for a rigorous and reasoned approach to holding in suspension two (or more) ideas that didn't appear compatible but which actually made sense for some aspect of our thinking (Stajduhar, Balneaves, and Thorne 2001). For me, philosophy was always more about the questions than the answers. I think I had long since become highly frustrated by the rapidity with which some nurse scholars sought definitive answers to what I took to be profoundly complex and difficult questions. The idea that answers were the goal seemed to me to be an artifact of the science within which we had been trying to force-fit ourselves since the post-WW2 era, when nursing began to aspire toward formal scientific credibility. So out of all of these frustrations, I found myself caught in a lifelong curiosity about how ideas work within a discipline like ours, and how we work with – rather than harness – those ideas to do what we aspire to do in a better way (Thorne 2007).

2. What, in your view, are the most interesting, important, or pressing problems in contemporary philosophy of nursing?

I think that the "identity politics" we see so prominently in theoretical discourse in nursing is the most pressing and worrisome of problems. Modern nursing is privileged to sit in the middle of and benefit from a dizzying array of great ideas from which to consider the questions that drive us. And yet I still see us as having problems with our alignment to certain ideas, potentially to the detriment of our ownership of our own discipline as an idea unto itself.

In my early philosophical disputations, I took issue with some of the conceptual model builders, particularly those nursing theorists whose overarching aim seemed to be the creation of dominance within disciplinary thought (Thorne et al. 1998; Thorne, Reimer Kirkham, and Henderson 1999). That competitive slant seemed to me a manifestation of the ethos of conventional science, in which it was traditionally assumed that one theoretical direction or paradigm would ultimately prove itself to be correct while others would fall by the wayside. I also took issue with various socio-political slants, such as feminisms, which could be wielded in the wrong hands as even more obnoxiously oppressive ideologies than the ideologies they sought to smash (Thorne and Varcoe 1988). I found it refreshing and exciting to "take aim" at ideas I had a deep personal attachment to (and still do) but to realize how valuable it was to also deconstruct them (Thorne 1999). I thoroughly enjoyed trying to dig beneath the surface of theoretical claims such as holism to try to discover the diverse ways in which people were employing and appropriating the concept toward various nefarious and divisive maneuvers (Thorne 2001). I was fascinated by how we use linguistic signals to cue one another as to which camp we favour, cite certain references and not others as authorities upon which we build, and bend these kinds of ideas toward the service of quite different motivational masters (Thorne et al. 2004). So ideas themselves really cannot be understood outside of the linguistic context of how we choose to use them, which takes me back to the fundamental problem I had with the hard-core concretizing and operationalizing concept work in the first place.

Another example of this kind of problem comes up in the world of social justice and action research in nursing (Thorne 1997). I fully understand the visceral appeal that various critical social theorizings, such as post-colonialism, neo-liberalism, even postmodernism have for many of my respected colleagues. But I believe that they really work best for those of us who work as nursing scholars as temporary standpoints rather than countries within which we take up citizenship. For me, the beauty of these newer ideas is that they inform and enlighten the discipline; however they tend to only do that if the holder actually treasures the ideational foundation of the discipline enough to cling to his or her original passport even when they do feel in a foreign land. So I see far too many nurses "becoming" something else as the desired endpoint of their intellectual adventures, trying to convince us all to go that particular direction rather than really coming home to layer their expanded consciousness into our own foundational platform. I am uncomfortable in a theoretical world in which the nursing becomes subsidiary to a scholarly trajectory as a critical social theorist, or feminist postcolonial scholar, and want to remind people what a marvelous kaleidoscope of thinking our own discipline really constitutes.

More recently I've taken that issue into the domain of qualitative research methodology. In my recent explorations, I have been challenging nursing to reclaim the epistemological structure of the discipline as the fundamental driver of design determinations, rather than perpetuating the trop of alignment with various theoretical directions invented for some alternative purpose that, when enacted beyond window dressing on our studies (and most often that is all they are), can seriously damage the intellectual line of logic that we take when we conduct a technically sound study (Thorne 2008). So I have been advocating the abandonment of the conventional "named" methods from the social science world and instead unpacking from their excellent repertoire of specific techniques and tools those that have particular relevance and promise within our own disciplinary logic model and toward our own intellectual purposes (Thorne 2011b). The techniques available to us within the established qualitative repertoire are splendid, but not if you are expected to subscribe to the religion in order to access them.

3. What, if any, practical and/or socio-political obligations follow from studying nursing from a philosophical perspective?

I think that any nurse who has had the privilege of indulging in the study of nursing from a philosophical perspective – whether through doctoral study or some other self-directed or alternative path – has an obligation to model for other nurses a relationship to ideas that invites curiosity. While we all imagined at the outset that our advanced education would help us know more, immersion in philosophy reminds us of the enormity of the domain within which we really don't and can't claim any solid knowledge. However, claiming to know less is also paradoxically consistent with finding more confidence in the world of ideas. We can't approximate wisdom until we know that there is always a great deal beyond our grasp. And we can't operate with that knowledge unless we become truly confident with the idea that our discipline is an enactment of ideas, ever-changing and transforming, always elusive, never stagnant, within the framework of a solid foundational core.

An example may illustrate. I do qualitative research, and I watch with horror the rhetoric around "theoretical saturation" that is routinely used by so many nurses who work in that genre, regardless of their methodological orientation (Thorne and Darbyshire 2005). The notion is a metaphor for a physics principle in which it does actually make sense; the porous object reaches a point at which it can no longer absorb, and the metaphorical diaper begins to leak. And in the theoretical sense, it says that no more new insight is possible even if I kept searching. So there we have it, some arbitrary way of determining that one's analysis is sufficiently deep, rigorous and elegant that it answers all possible problems and that no more new cases could possibly challenge it. While I believe that this idea might resonate intellectually if you were a social scientist and your point in life was to align with a dominant grand theory, I fundamentally and totally reject the notion that this idea that should have legitimacy within a nursing lexicon. The wisest nurses I know are those whose careers always and forever remain open to one more new layer of insight that they have not yet perceived, whose eyes and ears are always attuned to new angles from which to glean inputs that will enrich and enhance their capacity to approach the complexity and diversity of human health experience. So when a nurse makes an unqualified claim about theoretical saturation as a justification for withdrawing from data collection, he or she is ostensibly claiming to have created sufficient conceptual boxes within which to fit all possible configurations of experience with which patients might confront us. And from a nursing perspective, that kind of thinking seems quite dangerous. No matter how elegant and rigorous our theorizing, no matter how neat and tidy a conceptual structure might seem, I would hope that every nurse understands that a knowledge base such as ours is inherently and inevitably dynamic. There will always be new intersectional layers that might render visible that which was previously obscured. We will always be striving toward better and more comprehensive ways of apprehending our business and turning those apprehensions into meaningful and moral practices.

So I see all of us who work within the philosophical rubric as having an obligation to keep ideas moving, to ensure that the questions continue to be asked and that the discipline remains open to exposing new layers of inquiry. Today many of us are caught up in matters of profound concern for the social determinants of health and the health inequities they produce. This is all such important work, and clearly consistent with our profession's essential social mandate. But to presume that each new layer of our critical social theorizing will be the pinnacle of our thinking, or to discount the probability that the next generation will work out even more creative and interesting ways to think about our profession's mandate and moral obligation, would be shortsighted. Instead we should be encouraging the next iteration of thinkers to understand where we have been, where those of us in the position of expressing our ideas at this time in our history thought we were heading, and then stand back and try to follow in the directions they begin to lead. The science – and the philosophy – will be much more interesting and relevant if we are excited about that future instead of trying to carve monuments into the achievements we have made in the present.

4. In what ways does your work seek to contribute to philosophy of nursing?

At one time I would have considered my philosophical work and my methodological activities as two separate strands in my scholarly program of activity. However, as I learn and evolve, I am increasingly convinced that they are but two sides of the elephant offered to us by the axiomatic Sufi parable. For me, that elephant really is nursing. I am no longer worried, as so many of us seemed to be just a few decades ago, that nursing is difficult to delineate and define with any precision. Instead I rather like that, because it places nursing in the larger scheme of things up there with the really big and important abstractions that shape our human sensate universe – like humanity, dignity, God, spirituality, the environment, world peace.

I actually don't mean to be facetious here, because I do believe there is an essential nature to this discipline of ours that transcends time, space and context, that nursing has a coherent ontological nature, and that it is "knowable" and recognizable both objectively and subjectively. It is a "real" entity even as it sits epistemologically at the impossible end of the idealism spectrum. Because of this, it seems to me that nursing is inherently worthy of the very best minds society has to offer as we all seek to study and discover this fascinating phenomenon together and over time.

I have had the privilege of teaching graduate students for what is now a fairly lengthy academic career. And what I love about those classes is that nurses of all ages and stages, from every imaginable practice context, can come together in a dialogue and truly "know" one another. When one student recounts an example from his or her practice of moral distress or of clinical excellence, the others engage with it in an immediate recognition of the essential nature of the challenge, whether they have any technical knowledge of the activities being depicted or experiential familiarity with that kind of patient. We can "know" good or poor nursing across vast substantive gulfs. Our patients (and their loved ones) "know" good or poor nursing as well and their conceptualizations are quite compatible with our own. While I am not suggesting that the subjective experience on the receiving end is the only or most accurate measure, it is a deeply meaningful indicator nonetheless. I therefore feel compelled to think of nursing as if it does have a deeply coherent and comprehensive essential structure, even as I also acknowledge that the best of our minds have not yet been able to fully and comprehensively delineate that. And to me, that just adds to the excitement.

5. Where do you see the field of philosophy of nursing to be headed, including the prospects for progress regarding the issues you take to be most important?

I do see a new generation of nursing scholars well-grounded in their relevant science – and in this day and age that requires both the qualitative and the quantitative methodological arts. When they are well grounded, that means to me that they understand these matters well beyond technique and into the fundamental worldview that each of the operational approaches implies. Through an understanding of both the philosophy of science and the insights that can derive from the various critical standpoints from which we consider the core problems facing the profession, the next generation will be much better placed to strategically steer their science toward impact. Our generation has clearly demonstrated a strong capacity to "do science," but has not yet effectively aligned that with a robust capacity to shape policy. We tend as a profession to be rather naïve in the larger world of making things happen. While pockets of excellence are clearly evident, we haven't yet learned how to come together in politically astute and strategic ways to align the fruits of our science with the social action that will be necessary to solve the social and health problems that our profession "sees" in a different manner than does any other group within society.

I hope that nursing can play an active role in unpacking the evidence rhetoric with which health care – and to some extent the larger society in which it is contextualized – are infused. We know science both as a source of useful facts and also as a perpetual puzzle. Any expert nurse simultaneously practices that which science tells us is the best approach and also modifies practice according to a legitimate disciplinary understanding of appropriate modifications based on individual variations (Thorne and Sawatzky 2007). Our discipline may therefore be ideally placed to speak as the voice of reason in the warring discourses around what science contributes to public policy. We know full well that evidence claims serve policy makers well when convenient and are ignored when inconvenient. But we tend not to put our minds toward understanding the ingredients of those public policy decision-making processes or inserting ourselves into influencing them. If the next generation of nursing scholars took up the complex challenge of understanding human political processes the way our predecessors took up trying to understand the complexity of the individual patient, we might be able to build that more fulsome understanding of how the world works into our disciplinary foundations as core curriculum. We needed to understand medicine in order to serve patients who are under medical care, and I think we likely need to understand political processes in order to serve patients whose health status is so profoundly shaped by the dynamics of the worlds within which they reside. The "ancestors" – thoughtful nurses like Florence Nightingale, Ethel Johns, Lillian Wald, Lavinia Dock, and Margaret Sanger – all would have understood that in their time.

I also hope that we will continue to see the proliferation of opportunities to "philosophize" within nursing doctoral programs, as with all other levels of nursing education. From my perspective, at this juncture in our history, graduate education in nursing devoid of philosophizing becomes hollow and incomplete. We are living in an era in which numbers are driving so much of our decision making. There will naturally be ongoing internal and external pressures to expand the cadre of doctorally educated nurses more quickly and efficiently than is currently the case. We will, increasingly, bend to the call for accessibility and flexibility in educational models. Further, as doctoral education proliferates beyond the research-intensive universities and into a wider range of institutions, this accessibility and flexibility will play an ever more prominent role in successful marketing strategies to draw qualified candidates away from the more conventional seats of higher learning. There will be continued pressure to trim curricula and at the same time to expand interdisciplinarity in graduate education. In this context, I sincerely hope that we will remain mindful of program proliferation politics and ensure that the standard of learning in our profession – especially at the doctoral level – preserves the expectation that holders of the doctoral degree in nursing have had a healthy exposure to and engagement in disciplinary philosophizing.

When I entered the profession as a nursing student in the late 60s, the study of "nursing history" had become rather unfashionable. "Nursing process" was all the rage as a systematic and orderly way to enact the logical problem solving of the discipline. During my post-basic education in the 70s, I encountered "conceptual frameworks" (also known as "nursing models" or "nursing theories"). Although they initially struck me as a total absurdity – and I must have been an annoying student in those days – I eventually came to love them for what they were trying to tell us about complex adaptive phenomena and dynamic interaction. Despite categorically rejecting them as prescriptive entities, I saw them as delightful architectural models of the complex conceptualizing and iterative praxis that is the profession at its finest. Later on, through my own doctoral education and the opportunity to engage with models and frameworks through the fresh eyes of a couple of decades of wonderful graduate students, I came to understand them as highly creative attempts to philosophize the ontology and epistemology of nursing, all the more impressive for having done so without the benefit of the kinds of philosophical frameworks, such as complexity science, to which we have access today (Thorne 2011a). I understand, but deeply regret, that the study of the theoretical frameworks developed during that era of model building has almost vanished from most graduate curricula today, except for a few holdouts of the more proselytizing variety. To me, awkward and unwieldy as they may be, the conceptual models individually and collectively yield up wonderful insights as the documentation of an important part of our intellectual evolution. I therefore relish the prospect of a resurgence of interest in studying "the nursing theories" as a remarkable part of our discipline's philosophical history.

In conclusion, I do see nursing becoming more comfortable with its unique and particular intellectual tradition and affording it its rightful space in the worlds of both practice and ideas. As nurse scholars come to recognize the value of relinquishing an uncritical attachment to the divisive "isms" (holism, postmodernism, poststructuralism, feminism, postcolonialism) and "paradigms" (qualitative/quantitative, simultaneity/totality), they begin to discover that there is in fact a richness of intellectual fodder within the ideational structure of the discipline itself. I therefore look forward to a resurgence of enthusiasm for the kind of philosophizing that advances and integrates, rather than fractures, our disciplinary thinking. After all, no matter how hard it ever was to define this profession of ours, we have always had the capacity to "know" it.

References

Stajduhar, K. I., L. Balneaves, and S. E Thorne. 2001. "A Case for the "Middle Ground": Exploring the Tensions of Postmodern Thought in Nursing." *Nursing Philosophy* 2: 72-82.

Thorne, S.

1997. "Praxis in the Context of Nursing's Developing Inquiry." In *Nursing Praxis: Knowledge and Action*, edited by S. Thorne and V. Hayes, xi-xxiii. Thousand Oaks, CA: Sage.

1999. "Are Egalitarian Relationships a Desirable Ideal for Nursing?" *Western Journal of Nursing Research* 21: 16-34.

2001. "People and Their Parts: Deconstructing the Debates in Theorizing Nursing's Clients." *Nursing Philosophy* 2: 1-4.

2005. "Guest Editorial – Conceptualizing in Nursing: What's the Point?" *Journal of Advanced Nursing* 51: 107.

2007. "Conceptualizing the Purpose of Nursing: Philosophical Challenges in Creating Meaningful Theoretical Learning

Experiences." In *Teaching Nursing: Developing a Student-Centered Learning Environment*, edited by L. E. Young and B. Paterson, 347-363. Philadelphia, PA: Lippincott, Williams, and Wilkins.

2008. *Interpretive Description*. Walnut Creek, CA: Left Coast Press.

2011a. "Theoretical Issues in Nursing." In *Canadian Nursing: Issues and Perspectives*, edited by J. C. Ross-Kerr and M. J. Wood, 85-105. Toronto: Elsevier.

2011b. "Toward Methodological Emancipation in Applied Health Research." *Qualitative Health Research* 21: 443-453.

Thorne, S., C. Canam, S. Dahinten, W Hall, A. Henderson, and S. Kirkham. 1998. "Nursing's Metaparadigm Concepts: Disempacting the Debates." *Journal of Advanced Nursing* 27: 1257-1268.

Thorne, S., and P. Darbyshire. 2005. "Landmines in the Field: A Modest Proposal for Improving the Craft of Qualitative Health Research." *Qualitative Health Research* 15: 1105-1113.

Thorne, S., S. Reimer Kirkham, and A. Henderson. 1999. " Ideological Implications of Paradigm Discourse in Nursing Research, Education and Practice Theory." *Nursing Inquiry* 6: 123-131.

Thorne, S., and R. Sawatzky. 2007. "Particularizing the General: Challenges in Teaching the Structure of Evidence-based Nursing Practice." In *The Philosophy of Nursing Education*, edited by J. Drummond and P. Standish. New York: Palgrave Macmillan.

Thorne, S., and C. Varcoe. 1988. "The Tyranny of Feminist Methodology in Women's Health Research." *Heath Care for Women International* 19: 481-493.

Thorne, S. E, A. D. Henderson, G. I. McPherson, and B. K. Pesut. 2004. "The Problematic Allure of the Binary in Nursing Theoretical Discourse." *Nursing Philosophy* 5: 208-215.

24

Francine Wynn

Senior Lecturer, and Director of Undergraduate Programs

Lawrence S. Bloomberg Faculty of Nursing

University of Toronto, Toronto, Canada

---

1. How were you initially drawn to philosophical issues regarding nursing?

I am a latecomer to philosophy; I see myself as brokering or mediating between specific philosophical works from contemporary continental philosophy and particular concerns arising in nursing practice. My study of philosophical texts has always been grounded in my nursing work and my perspective would be very different if I had studied philosophy before becoming a nurse. On the one hand, engagement with philosophy creates dilemmas for me, as it is very difficult to keep abreast of current discourse in philosophy and requires time and space to read and think. On the other hand, we nurses have actual concerns to address arising from the care of suffering persons that marks our projects as different from traditional philosophical work.

I was 'trained' to be a nurse in the mid-sixties in a hospital program in Montreal and this apprenticeship shaped all my future endeavours in nursing and philosophy. 'Being trained' offered a unique perspective on nursing and the medical world that cannot be replicated in nursing education today. My everyday life as a student was immersed in the hospital world as I, along with my peers, lived on the hospital grounds at first in a residence attached to the hospital and latterly in a huge old house nearby. I ate all my meals in the hospital cafeteria, as we never had access to a kitchen in which to prepare food.

This was a strange world and was endlessly fascinating to me. From the very beginning I was both drawn to and deeply disturbed by the drama and the suffering I encountered within the institution. I have always been most interested in the particular patients I was caring for and their situation. I was not particularly drawn to the anatomical-physiological science of the body but rather the fleshy person in front of me. The on goings of the medical environment in all of its technological complexities fascinated yet disturbed me. I clearly remember my first dying patient – a man suffering with oesophageal varices. Immediately after his death the resident exclaimed "now we have a bed to admit to." I was horrified by this seeming indifference. This response by the resident I would now call a deficient mode of care or solicitude, following Heidegger, and preoccupation with the importance of 'caring' for the sick drives much of my philosophical writing.

I was always dissatisfied with how we were taught in this apprentice system, the lack of depth in the knowledge we received and the passivity that was systematically encouraged. I was acutely aware of the tension between needing to know and learning how to practice safely, and the 'service' we were required to perform. We students were the nursing labour that kept the institution going and our studies were secondary. It was not unusual to come off of a night shift and have to attend class. On the other hand, the hospital was a huge learning laboratory that offered endless encounters with new situations. Understanding the rules and rituals of the hospital was a necessity for getting the work done but was learned in practice not in class.

I practiced in a small coronary care unit after graduation in my home city but found it distressing to care for people whom I knew. I left after three months and began working as a staff nurse in a university psychiatric institute in the new field of community psychiatry. This work was very exciting, innovative and intellectually stimulating. Questioning and active discussion within the team was the daily stuff of clinical work. I was aware from the beginning that I needed university preparation to deepen my practice. After four years in two other psychiatric settings I went to university full time. Anthropology seemed an ideal discipline through which to explore my curiosity about what shapes persons and contexts and I studied for 5 years working two evening shifts every weekend.

From the outset I have to clarify that I was not drawn to philosophy *per se* but by a philosophical way of approaching and interpreting situations that can be generally subsumed under the rubric of interpretive phenomenology. The anthropologist Clifford Geertz, for example, was very important for me. I only discovered continental philosophy after completing two degrees in socio-cultural anthropology and a year of a PhD in medical anthropology. At the time I audited a course in contemporary social theory in which we read Benjamin, Habermas, Adorno, Derrida, Foucault, Merleau-Ponty, Baudrillard etc. I joined the Program in Social and Political Thought at this point to pursue theory exclusively rather than do fieldwork, as I felt that my grounding in anthropological theory needed to be supplemented if I was to make sense of what I was encountering in practice and in teaching. I read widely in my program and originally intended to work on Foucault. After completing my coursework I audited a philosophy course of Sam Mallin's on late Heidegger and Merleau-Ponty; since then my philosophical approach draws almost entirely from the writings of Heidegger, Merleau-Ponty, Agamben and Nancy.

2. What, in your view, are the most interesting, important, or pressing problems in contemporary philosophy of nursing?

From my perspective one of the most pressing problems is the 'apparent' demise of relational work at the bedside. In acute care settings in North America and elsewhere, relationally grounded bedside nursing seems to be on the verge of extinction apparently due to severe (nursing) shortages, the increasing complexity of the work due to the high acuity of patients and short hospital stays, the focus on populations of patients rather than individuals, and high treatment costs. There appears to be little time to do basic and urgent patient care, much less affective work. Registered nurses at the bedside are being replaced by unskilled workers and nursing work is shifting to distally managing patient through input and flow, information, resources and orchestrating the work of doctors and other health professionals.

How might we understand this shift philosophically? How are patients encountered in what Heidegger (1977) calls the epoch of technicity in which patients potentially are merely resources to be shifted about? He describes this regime as an aggressive-unlocking challenging-out that systematizes everything. All beings (human, natural and constructed) become overseeable, controllable, definable, connectable, explicable, stockable, and transformable (Heidegger, 14-16). The emphasis is on acceleration and efficiency. As a result, nature is understood as a mere resource waiting to be exploited for its energy, rather than as having a rhythm and order of its own. Modern science, under this ruling, reveals beings as already prior-determined object-representations, characterized by exactitude, and posited or projected in advance as knowable and researchable. Everything is seized and submitted to a systematic ordering that is thoroughly accessible to instrumental measure and calculation.

What is the value of relational work under such a regime and why should we defend its necessity philosophically? My method is grounded in the description of ways in which things, people and relations come into appearance. Of particular interest to me is our embodied relation with others and with technology. In this phenomenological approach, what currently comes into appearance is the regularization and systemization that is part of contemporary nursing practices. For example, we in nursing seem caught up in patients always being the same. In every encounter we seem to want or expect the same identity to exert itself. Jean Luc Nancy's understanding of singularity, for instance, might help free us in situations that trouble us because the patient seems difficult or noncompliant or unwilling to accept what we want to do to him or teach him etc. In such situations nurses seem to avoid encountering the patient at all, and when they do approach him they are not attentive to him but rather brusquely get in and get out of the room as fast as possible.

My original philosophical work (Wynn 1996, 1997, 2002), drawing on the phenomenology of Merleau-Ponty, is a description of the inter-corporeal mother-infant relation understood as a hidden chiasmic holding, sheltering, co-relating, in which not only the mother actively brings the child into the world, but the child transforms the mother. I argue that this holding-being held of mother and infant grounds both of their future perceptions and inter-corporeal relationships. This approach can be brought to an understanding of the patient-nurse relation as a potentially non-dominating relation in which nurse and patient mutually shape the other. In addition both can be seen to be affected by earlier holding relationships that have been sedimented bodily but not known self-consciously.

In a later work, "Art as Measure" (Wynn 2006), I draw on Heidegger's thinking about our primordial dwelling (Heidegger 1971) to open up nursing as a cultivating safeguarding which preserves and protects the patient. Here I differentiate between a necessary patient-safety protocol-driven practice in which safety seems to be driven by a snatching from danger, and a more originary safeguarding characterized by a co-operative attentiveness that shares both agency and final outcome and frequently is inconspicuous. But nursing now is being shaped more and more by systematizing rather than relations with particular patients who have specific concerns. Contact with patients is decreasing and becoming more fragmented. Systematicity covers over these relations. Are we currently in a situation in which this ever-increasing acceleration will obliterate this relational safeguarding? Is this relation capable of being erased by technicity and can we resist this? How are we to think through the importance of relational nursing work at the same time that it seems to be disappearing?

Stimulated by the opportunity this publication offers and the necessity to consult specific philosophical texts in such an enterprise, I will try to think through, in a very preliminary and attenuated way, some of Heidegger's ideas about the ontology of Dasein as being-with (Mitsein). My philosophical approach is to work with specific texts, much as I work with the particularity of patients and students. This kind of reading is an alongside reading where questions in the texts help us to open up questions of practice and vice versa. I first turn to Heidegger's

Zollikon Seminars (Heidegger 2001) to think again about this question of human relations. I start here for two reasons. First, in these seminars Heidegger shares the relevance of phenomenology for the practice of medicine with a group of psychiatrists. Second these seminars, given over a period of 10 years between 1959 and 1969, give access to Heidegger's late thinking in which his critique of technicity occurs together with his phenomenological examination of Dasein as being-in-the-world. It becomes apparent in reading these seminars that he was offering these psychiatrists a "crash course in concepts from Being-and Time" (Richardson 1993). His discussion of the phenomenon of being-with is scattered throughout the seminars. He declares: "In *Being and Time* it is said that Dasein is essentially an issue for itself. At the same time, Dasein is defined as originary being-with-one-another. Therefore, Dasein is also always concerned with others. Thus, the analytic of Dasein has nothing whatsoever to do with solipsism or subjectivism" (Heidegger 2001, 120). As Heidegger is staying close here to his original phenomenological examination of being-with I decide to go directly to Section 26 in *Being and Time* (1962) in which he examines Dasein as being-with in more depth and detail. I open up his concepts of positive and indifferent modes of solicitude, which are grounded in working alongside others in their circumspective concern with equipment in the ready-to-hand.

Here, I must point out that I am not reading Section 26 in a general sort of way but bringing my nursing queries with me to the text. I am questioning the meaning of bedside nursing work and its grounding as relational rather than merely technical skill. In particular I am struggling with situations culled from beginning nursing students' descriptions of observing the postpartum nursing care of mothers and newborns, and elderly patients in long term care. In these observations mothers of newborns are being approached as feeding machines and elderly patients tended to be cared for as things rather than persons. How might a return to the phenomenon of 'being-with' help us resist caring for mothers and babies and the elderly in this way? Currently postpartum care is brief as mothers and their infants are being discharged within 24 – 36 hours. Nursing practice seems driven by specific protocols that attempt to ensure that mothers are taught how to breastfeed correctly before they leave. This is a speeded-up process that focuses on an infant's latching properly onto mother's nipple. Breastfeeding is seen as the best possible practice and its nutritional rather than relational qualities are promoted. Mothers seem quite anxious about their abilities to successfully nurse and breastfeeding clinics both public and private are popping up due to the difficulty mothers are having breastfeeding when discharged from hospital. How might a return to the phenomenon of 'being-with' as described by Heidegger open up these disturbing situations?

Heidegger claims that both caring for things and caring for Others are to be interpreted under the more primordial phenomenon of 'care'. These all are ontological properties of Dasein. In section 26 he specifically analyses the relational mode of the 'with' of the being-with-others. He distinguishes between a caring for things that he calls circumspective concern (Besorge) and a taking care of or caring for others as different types of comportment he calls solicitude (Fuersorge), although both can share the characteristics of "inconspicuousness and obviousness" (Heidegger 1962, 158). English is a problem here as the root word for care is in German is *sorge*. On the one hand, circumspective concern is a kind of concern which "manipulates things and puts them to use" (Heidegger, 95) and is a way of "discovering what is ready-to-hand" (Heidegger, 159). On the other hand, although Others are often encountered in terms of what is ready-to-hand, as we encounter them at work, they are not Things present-at-hand. Dasein-with is an essential structure and Others always are "already there with us;" "a bare subject without a world never is proximally, nor is it ever given" (Heidegger, 152). But how does this already there with us come about?

To quickly summarize, Heidegger approaches Dasein's 'being-with' in its average everyday understanding. For the most part he claims that being-with is, for the most part, carried out in a deficient or indifferent mode of solicitude (fuersorge), arising in working alongside others in their circumspective concern (besorge), with things/equipment objects in the ready-to hand. Only in certain circumstances do positive modes of solicitude arise, which he characterizes as 'leaping in' and 'leaping ahead'. These positive modes have been examined in nursing most prominently by Benner and Wrubel (1989, 48-49) and Scudder (1990). I am expanding here on their work.

Heidegger opens up his analyses of solicitude with the statement: "Even 'concern' with food and clothing, and the nursing of the sick body, are forms of solicitude" (Heidegger 1962, 158). This assertion could be taken at face value to be potentially about nursing care. The word 'even' sets up a kind of curious tension as if the preparing of food and perhaps feeding of others, the making of clothes and the dressing of others and nursing the sick body are unlikely candidates for solicitude. It must be noted here that the word for body here is *Leib* rather than *Koerper*. Is this due to Heidegger's ambivalence towards such relations or his focus on the inauthentic of the everyday? What then follows in this rather brief paragraph is a discussion of the more predominant negative-privative or deficient modes of solicitude rather than the positive modes. Heidegger specifically addresses 'welfare work' as a type of work that is about feeding and clothing and nursing the poor. He remarks that such work's "factical urgency" gets it motivation in that Dasein maintains itself proximally and for the most part in the deficient modes of solicitude (Heidegger 1962, 158). This seems to indicate that such needs of the vulnerable are overlooked because average everydayness of Being-with-one-another for the most part is based in indifference, such as: "Being for, or against, or without one another, passing one another by, not "mattering" to one another." Such indifferent modes are likely a necessary way of being-in-the-world as we ride subways, buy items in stores, listen to lectures, attend football games, engage in machine work etc. In such indifferent and deficient modes we tend to push others away. But is being-with in the nursing of the sick in its average everydayness necessarily based in indifference?

Next Heidegger turns to his analyses of the positive modes of solicitude in which we are drawn towards others in their particularity rather than encounter them in the "distance and reserve" of deficient and indifferent modes (1962, 159). It is helpful to keep food and clothing and nursing of the sick body in the foreground when reading his opening up of "the two extreme possibilities of the positive modes of solicitude" (159). The first extreme mode 'leaps in' for the Other; "it **can** take care away from the Other and put itself in his position in concern" (1962, 158). As a consequence the Other is thrown out of his position. This type of solicitude is a 'taking over' and the other "**can become** one who is dominated and dependent, even if this domination is a tacit one and remains hidden from him" (158). If we turn to the example of feeding, dressing, and nursing of the sick we can think through this extreme mode of solicitude. In this situation the Other is unable at the present to follow through with his concerns (besorge). Perhaps the elderly patient is too confused or too weak to feed himself or dress himself or get up from bed or carry on with his projects in his own manner of getting on with his daily routine, or the new Mother is not yet able to feed her baby in her own way or according to the baby's style as she has just given birth and her baby is so new she has no confidence in how to hold or feed her etc. What does it mean to be thrown out of one's position? For Heidegger this means that one's ready-to-hand involvement in a familiar practical world is being disrupted by an illness or a new event. Maybe the nurse steps in and takes over the feeding, dressing, washing, nursing etc. Does she necessarily have to dominate and make the Other dependent in doing this? In the student observations of breastfeeding and care of the elderly it appears that it is the nurse who decides breastfeeding (based in evidence based protocols) is best regardless of the feelings of the new mother, or that the elder person will be fed quickly rather than helped to feed himself. Yet, in healthcare settings patients do come to us in vulnerability and potential dependence and we do need to leap in and take care away from them **temporarily**. Such care arises in situations as something necessary but not indefinitely so. If we take care away in such a manner that the Other is unable or unwilling to take it over herself, then this likely will lead to dependency and domination. Heidegger contends that leaping in is a much more common kind of solicitude and "is to a large extent determinative of our Being with one another, and pertains for the most part to our concern with the ready to hand" (Heidegger 1962, 158). At this point in the text Heidegger does not comment on the style of the care-taker when leaping in but remarks in a brief paragraph on the next page that solicitude can be guided by *considerateness* and *forebearance* or *inconsiderateness or perfuntoriness* (Heidegger 1962, 159). It seems that in our two clinical examples, drawn from student nurses' observations of mother and babies and the elderly, perfunctoriness rules. Leaping in may not be the most predominate mode of solicitude at all, but rather, these nursing inter-actions, at best, are comported in indifference.

Heidegger then turns to the other extreme possibility of positive solicitude he calls "leaping ahead" or "leaping forth" that comports itself "not in order to take away his 'care' (sorge) but rather to give it back to him authentically as for the first time" (Heidegger 1962, 159). Note that the term in German is sorge not besorgen. Here what is at stake is not the Other's concern, his 'what', but his 'care' (sorge) – what is authentic or proper to him and is focused on a future potentiality. "It helps the Other to become transparent to himself *in* his care (sorge) and to become *free for* it" (Heidegger 1962, 159). In leaping ahead there seems to be an assumption that the carer has knowledge of the situation of the Other, yet doesn't direct her and take over but is able to guide or coach the other to become transparent to herself about what her life and projects mean to her. He amplifies that this stance that encourages self-transparency or "self-disclosure" "grows only out of one's primarily Being with him in **each** case" (Heidegger 1962, 161). So to return to our mothers and babies, how might the nurse proceed in such transient circumstances? How might she guide the mother to come to her own understanding and decision about how she will feed her newborn? How might she foster the relational and nutritional qualities in breastfeeding without dominating? Doesn't breastfeeding open up the very meaning of mothering?

It must be pointed out that for Heidegger, being-with arises from **common** concerns in the everyday work world of equipment. Dasein inhabits a practical world and everything we deal with is equipment. For Heidegger, Being-with does not have its origin in solicitous relations *per se*, but is grounded in our everyday relations around our common concern with entities/things. He states: "Solicitude proves to be a state of Dasein's Being- one which, in accordance with its different possibilities, is bound up with its Being towards the world of its concern, and likewise with its authentic Being towards itself" (Heidegger 1962, 159). Heidegger remarks that to describe and classify the positive modes and its numerous mixed forms to be "beyond the limit of this investigation" (Heidegger 1962, 159). He asserts that Being-with one another is "often exclusively based upon a matter of common concern.... arising from one's doing the same thing as someone else.... in a mode of distance and reserve and often thrives only on mistrust" (Heidegger 1962, 159). However there is another possibility that arises if they "devote themselves to the same affair in common, and their doing so is determined by the manner in which their Dasein has been taken hold of. They [thus] become *authentically* bound together, and this makes possible the right kind of objectivity, which frees the Other in his freedom for himself." Yet is breastfeeding the same affair in common for both the nurse and the mother? Is breastfeeding merely based in the ready-to-hand as if the breast were an object or device? Or, must we search for another understanding of being-with that has its origins elsewhere.

Here I can only touch briefly on such an alternative arising in the philosophical writings of Luce Irigaray, specifically *Sharing the World* (2008), which I have yet to work through. Although deeply indebted to Heidegger, Irigaray takes his analyses further. She asserts that his understanding of being-with is a technological one. She finds Heidegger guilty of a masculine reading that assumes everyone inhabits a single common world. She troubles this notion of an unquestioned shared common world of equipment and the ready-to-hand as the basis of being-with. She suggests that the origins of being-with instead arise in the child's forgotten encounter with the mother's world and her care (fuersorge). She claims: "being with, or to, the other is always already constitutive of our existence and our way of being in the world. The first world with which the subject has to deal is the other. The first world of the subject is an other. The little human lives in a world that is the world of its mother" (Irigaray 2008, 105). Her thinking helps us appreciate the necessity of nursing relations with the Other based in a reinvigorated everyday positive solicitude originally grounded in maternal care. Unfortunately, the articulation of this necessary positive solicitude is a project for another day.

3. What, if any, practical and/or socio-political obligations follow from studying nursing from a philosophical perspective?

My philosophical perspective is an essential part of my teaching and my work with faculty colleagues. My obligation is to try to stay awake! This is meant in a phenomenological way. I feel responsible to pass along as best I can what I have learned and continue to learn through my engagement with continental philosophy. I keep questioning the on-goings of technicity by pointing out to students and faculty how we are being systematized and unlocked by protocols and audits. I try to be open bodily to what is happening around me and try to resist being completely overtaken by the digital world we live in. I cultivate my perceptual capacities of listening and seeing and moving. For example, a few years ago I experimented with walking to work after seeing/hearing this amazing film about Evelyn Glennie called *Touch the Sound* (Riedelsheimer 2004). After viewing the film I heard sounds that I have never attend to before. My whole attention was sound-focused. It was a thrilling experience. Over a period of weeks I tried to listen by being attuned, alert and vigilant as I daily navigated my way to work. Rather than being caught up in the noises of cars, trucks, and streetcars, I was able to hear the sounds of birds, the rustling of leaves, and my own footfall. The walks vibrated in a new way. This experiment was shared with doctoral students and became the basis for an assignment.

I try to bring a chiasmic relating to my students and to the papers and texts we are reading. I encourage a close reading of original theoretical material if this is possible. I am always surprised what students can glean from original texts. I try to resist the demands to water down ideas or make them into bullet points. Teaching now is more and more an immobilized reading behind a lectern set up at the side of the class. The classroom is dark so that the powerpoints can be read, and lectures are demanded beforehand as if nothing new could happen in the class. It is much harder to actively engage with students in immediate lively way. I attempt to resist this, of course. With the hyper speed up of everything and the relentless drive for more information, our time for reading and thinking and our capacity to attend to the deep is being compromised. My greatest worry is there will be no time in undergraduate and graduate curriculums to pause and struggle with difficult ideas or appreciate the possible poetics of practice.

4. In what ways does your work seek to contribute to philosophy of nursing?

My work is a small attempt to bring certain key philosophical readings and concepts to specific nursing or medical situations. I am drawn to making visible or unconcealing the effects of technicity and returning nursing to the care of the sick. I also am committed to engaging with difficult philosophical ideas and to bringing them forward to a nursing audience. I feel strongly that nurses should be engaged in current philosophical discussions and that our practice has much to contribute to philosophical work. Armed with these ideas we can potentially change practice or at least create disruptions with the new insights we have gained. Examples from my work include writings on Heidegger's concepts of gestell/technicity and dwelling (Wynn 2006), Agamben's concept of bare life (Wynn 2002), Nancy's exploration of transplantation (Wynn 2009) and Merleau-Ponty's notion of the chiasm (Wynn 1996, 2002).

5. Where do you see the field of philosophy of nursing to be headed, including prospects for progress regarding issues you take to be important?

For 30 years I have brought cinema, poetry, and the plastic arts into the classroom to enhance and enliven the clinical appreciation of the rich texture of human existence. I have been encouraged to be creative in the classroom. The way things are going now I fear that more and more information and device-centred course work will be inserted into nursing curriculum and less time will be permitted to explore complex ideas either in the classroom or in assignments. Our fascination with simulation labs now seems to be overriding the necessity for clinical practice with actual patients. Now multiple choice exams are the evaluation method of choice. Nursing doctoral programs are four years in length with the focus on method rather than theory. Students will not have the opportunity to study philosophy in any depth. From a Heideggerian- based phenomenology, the belief in progress is a troubled one. He would see what is currently happening as a last man situation or the "end of philosophy" which is not its finish but its endless reiteration. He writes: "The end of philosophy proves to be the triumph of the manipulable arrangement of a scientific-technological world and of the social order proper to this world. The end of philosophy means the beginning of the world civilization based upon Western European thinking" (Heidegger 1972, 59). For him this is deeply problematic and the great danger.

Publications

Benner, Patricia and Judith Wrubel. 1989. *The Primacy of Caring*. Menlo Park, California: Addison-Wesley Publishing.

Heidegger, Martin. 1962. *Being and Time*. Translated by John

Macquarrie and Edward Robinson. New York: Harper & Row.

Heidegger, Martin. 1972. "The End of Philosophy and the Task of Thinking." In *On Time and Being*. Translated by Joan Stambaugh, 55-74, Chicago: University of Chicago Press.

Heidegger, Martin. 1971. "Building Dwelling Thinking." In *Poetry Language Thought*. Translated by Albert Hofstadter, 145-161, New York: Harper & Row.

Heidegger, Martin. 1977. "The Question Concerning Technology." In *The Question Concerning Technology and Other Essays*, Translated and introduced by William Lovett, 3-35. New York: Harper Torchbooks.

Heidegger, Martin. 2001. *Zollikon Seminars*. Edited by Medard Boss, translated by Franz Mayr and Richard Askay. Evanston Illinois: Northwestern University Press.

Irigaray, Luce. 2008. *Sharing The World*. New York: Continuum.

Riedelsheimer, Thomas. 2004. *Touch the Sound: A Sound Journey with Evelyn Glennie,* Germany.

Richardson, W. J. 1993. "Heidegger Among the Doctors" in *Reading Heidegger: Commemorations*, edited by J. Sallis, 49-63. Bloomington: Indiana University Press.

Scudder, James. 1990. "Dependent and Authentic Care." In *The Caring Imperative in Nursing Education,* edited by M Leininger and J Watson, 59-66. New York: NLN Publication

Wynn, Francine.

1996. *The Early Mother-Infant Relationship: Holding and Being Held*. Doctoral Dissertation, Toronto: York University.

1997."The Embodied Chiasmic Relationship of Mother and Infant." *Human Studies* 20: 253-270.

2002."Nursing and the Concept of Life: Towards an Ethics of Testimony." *Nursing Philosophy* 3: 120-132.

2002."The Early Relationship of Mother and Pre-infant: Merleau-Ponty and Pregnancy." *Nursing Philosophy* 3: 4-14.

2006."Art as Measure: Nursing as Safeguarding." *Nursing Philosophy* 7: 36-44.

2009."Reflecting on the Ongoing Aftermath of Transplantation: Jean Luc Nancy's *L"Intrus*." *Nursing Inquiry* 16: 3-9.

About the Editors

Anette Forss is a Senior Lecturer, Division of Nursing Karolinska Institutet, Sweden and a Visiting Scholar at the Department of Philosophy, Stony Brook University, SUNY (New York), USA. Her interest is to merge philosophy of technology (postphenomenology) and nursing. She is the principal investigator of an ethnographic study that explores the nurse-technology-patient interface in a variety of oncology related health care regimes. Besides *Philosophy of Nursing - 5 Questions*, she has also co-authored papers published in a variety of scientific journals and has authored a chapter in *Medical Technologies and the Lifeworld*, Routledge (2007).

Christine Ceci is an Associate Professor in the Faculty of Nursing, University of Alberta, Canada. Taking up questions of knowledge, power and organization, the current site of her research is supportive home care for those who are older and frail, in particular, the conditions of possibility that structure these practices. She is co-editor of *Perspectives on Care at Home for Older People* (Routledge 2011), an international, multidisciplinary collection of papers on the topic. She is also an active contributor to the work of the Institute of Philosophical Nursing Research (now the unit for Philosophical Nursing Research (uPNR)) at the U of A.

John S Drummond is a Senior Lecturer in Nursing at the University of Dundee, UK. Besides *Philosophy of Nursing - 5 Questions*, he also co-edited *The Philosophy of Nurse Education* with Paul Standish, Palgrave Macmillan, Houndmills, Basingstoke, and New York (2007). He is a co-founder and current Treasurer of the International Philosophy of Nursing Society (IPONS). He is also Book Review Editor (rest of the world) for the journal *Nursing Philosophy*, the official journal of IPONS, published by Wiley-Blackwell.

Index

E

F

G

H

I

K

L

M

N

O

P

Q

R

S

CPSIA information can be obtained
at www.ICGtesting.com
Printed in the USA
LVHW012243220922
729056LV00002B/243

9 788792 130495